*The Measurement
of Time-Varying
Phenomena*

The Measurement of Time-Varying Phenomena

Fundamentals and Applications

EDWARD B. MAGRAB

The Catholic University of America

DONALD S. BLOMQUIST

The Catholic University of America

WILEY-INTERSCIENCE, a Division of John Wiley & Sons, Inc.

New York · London · Sydney · Toronto

PREFACE

For many years the measurement of time-varying phenomena has been primarily the concern of a relatively small group of engineers and scientists. In recent years, however, other engineers, scientists, and medical doctors have become increasingly involved in measuring, recording, and analyzing time-varying information. Thus many people from various disciplines and backgrounds have had to use unfamiliar instruments and measurement techniques.

Overcoming the initial and fundamental problems that may be encountered by the neophyte usually requires both classroom instruction and laboratory experience. There is no substitute for the laboratory experience. However, the problems encountered by the beginner most often are concerned with the violation of the fundamental laws common to most instruments. Although most equipment comes with adequate instructions for its proper use, much of this information cannot be fully understood and appreciated without some fundamental understanding of the equipment.

The purpose of this book is to discuss in a single source the principles of operation and the limitations of the most commonly used instruments and measurement techniques of time-varying phenomena. The primary concern here is with the qualitative behavior of classes of instruments from an input-output point of view rather than with the actual details of construction. The book is intended for the instrument user rather than for the instrument designer. Thus the reader may have almost any engineering or science background, for the intricacies of complex circuit analysis are bypassed.

The book attempts to be reasonably self-contained, with some prior knowledge of calculus and elementary circuit theory assumed. The first chapter, although brief, introduces all the mathematics and definitions needed for a detailed understanding of the chapters following. The next six chapters discuss, both analytically and descriptively, the functions, properties, and limitations of the most common types of instruments used in the measurement and analysis of time-varying phenomena. They are, respectively, filters, amplifiers, voltage detectors, recorders, signal generators, and digital systems—components and techniques. These chapters are ordered so that each relies on some of the material presented in preceding chapters. This presentation leads to a "systems" approach in which the reader learns to piece together his own measurement system from knowing the function, limitation, error, and interaction each subsystem has. Discussion of transducers and the phenomena being measured are given in the applications. However, a detailed discussion of the transducer per se is not presented. Instead it is treated as a representative of a class of transducers having similar electrical properties.

The impetus for the book came from material used in one senior-level undergraduate course and two first-year graduate courses taught by us during the last four years in the School of Engineering at The Catholic University of America, which were entitled "Dynamic Measurements," "Fundamentals of Instrumentation," and "Experimental Dynamics Laboratory." These materials were put into their present form for a series of short courses entitled "Dynamic Measurements in the Ocean Sciences," sponsored by the Instrument Society of America and given four times during 1968 and 1969.

We acknowledge and express our appreciation to a number of manufacturers of scientific instrumentation whose engineers have produced abundant and well-written literature that has added to our understanding of instrumentation. They are Analog Devices, Inc., Norwood, Mass.; B and K Instruments, Inc., Cleveland, Ohio; General Radio Company, West Concord, Mass.; Hewlett Packard Company, Palo Alto, Calif.; Honeywell, Inc., Test Instruments Division, Denver, Colo.; Tektronix, Inc., Beaverton, Ore. In addition, we thank the Instrument Society of America for their support which made the early editions of this book possible. We are also indebted to Mrs. Mary Kurylo for her expert typing of the many stages of the manuscript, including the final one, and to Christopher Thomas for the art work. Last, but not least, are the contributions of the many students and friends whose comments and criticisms have proved invaluable. In this regard we especially thank John Wooten of B and K Instruments, Cleveland, Ohio, for his thorough reading and constructive criticisms of the original manuscript.

EDWARD B. MAGRAB
DONALD S. BLOMQUIST

Washington, D.C.
February 1971

CONTENTS

APPLICATIONS

The Measurement
of Time-Varying
Phenomena

1

ANALYSIS OF SIGNALS

1-1 Introduction

For purposes herein the following definition of a signal is employed: the variation through time of any significant physical quantity occurring in a useful device or system. Perhaps the most apparent feature of a signal is its wave form, that is, simply the graphical depiction of the signal as a time-varying quantity. As seen later the wave form of a signal is only one way in which a signal may be described.

To analyze signals it is often necessary to represent them by a mathematical function of time, or a group of such functions, each defined over a specified time interval. One of the most useful mathematical tools for the analysis of signals is harmonic analysis. Harmonic analysis employs Fourier series and Fourier transforms. Although harmonic analysis is capable of extension in many directions, this chapter is concerned with presenting only that portion of harmonic analysis necessary for an adequate understanding of the concepts involved in measuring time-varying phenomena. In addition, since the measurement of time-varying phenomena often involves, either by choice or by chance, nondeterministic or random processes, a brief treatment of probability theory and how it relates to the Fourier methods is presented.

1-2 Classification of Signals

Observed signals representing physical phenomena can be classified as being either deterministic or nondeterministic. Deterministic signals are those that can be described by an explicit mathematical relationship as a function of time. Nondeterministic signals cannot be so described and are usually termed random.

Deterministic signals can be further classified as either periodic or aperiodic (sometimes called transient). A periodic signal is one for which its shape (amplitude) as a function of time repeats itself in its entirety every time interval, or period, T_0. An aperiodic or transient signal is every other type of deterministic signal.

1

Nondeterministic or random signals can be further classified as either stationary or nonstationary. However, their precise definition requires the introduction of several terms which themselves must be defined. Consequently, their definitions are not given here and instead are stated in Section 1-5.

1-3   Periodic Signals

1-3.1   Fourier Series

A signal $g(t)$ is periodic on some interval T if $g(t) = g(t + T)$. The frequency at which $g(t)$ is periodic is given by $f = 1/T$ denoted by hertz. This frequency is related to the circular frequency ω by $\omega = 2\pi f$; thus $\omega = 2\pi/T$ rad/sec. Practically all physically realizable periodic signals can be represented by a properly weighted combination of two basic wave forms; the sine and cosine functions. These trigonometric functions have a period of 2π; that is, they naturally repeat themselves every $2n\pi$ radians ($n = \pm1, \pm2, \pm3, \ldots$).

Such a suitably weighted combination is given by

$$g(t) = \frac{a_0}{2} + \sum_{n=1}^{\infty} \{a_n \cos n\omega_0 t + b_n \sin n\omega_0 t\} \tag{1-1}$$

where the coefficients a_n and b_n are determined from

$$a_n = \frac{2}{T} \int_{-T/2}^{T/2} g(t) \cos n\omega_0 t \, dt \qquad n = 0, 1, 2, \ldots \tag{1-2a}$$

$$b_n = \frac{2}{T} \int_{-T/2}^{T/2} g(t) \sin n\omega_0 t \, dt \qquad n = 1, 2, 3, \ldots \tag{1-2b}$$

and $\omega_0 = 2\pi/T$. The quantity $\omega_0 = 2\pi f_0$ is called the fundamental frequency. It is seen that this representation of a periodic signal is a function of the coefficients a_n and b_n and the integer multiples of the fundamental frequency. These integer multiples are called harmonics of the fundamental frequency ω_0.

A series expansion of the form given by (1-1) and (1-2a, b) is called a *Fourier series*. This series is indispensable in the analysis of signals. It can be rewritten in another, and for our purposes a more useful, form. If $j = \sqrt{-1}$, then

$$g(t) = \sum_{n=-\infty}^{\infty} C_n e^{jn\omega_0 t} \tag{1-3a}$$

$$= C_0 + 2 \sum_{n=1}^{\infty} |C_n| \cos (n\omega_0 t - \varphi_n) \tag{1-3b}$$

where

$$C_n = \frac{1}{T} \int_{-T/2}^{T/2} g(t) e^{-jn\omega_0 t} \, dt \tag{1-4}$$

To determine C_{-n} we let $n = -n$ in (1-4). It is seen that, in general, C_n is a complex member whose absolute value is given by

$$C_n C_{-n} = |C_n|^2 = \frac{a_n^2 + b_n^2}{4} \qquad (1\text{-}5a)$$

where the a_n and b_n are those defined by (1-2a, b) and we have used the fact that C_{-n} is the complex conjugate of C_n. The angle φ_n is given by

$$\varphi_n = \tan^{-1} \frac{b_n}{a_n} \qquad (1\text{-}5b)$$

It is clearly seen from the form of (1-3b) that a periodic signal may be reconstructed by the sum of suitably weighted and phased (with respect to $t = 0$) sinusoids, the frequency of each sinusoid being an integer multiple of the fundamental frequency of $g(t)$. The quantity $|C_n|$ is called the amplitude spectrum of $g(t)$ and φ_n its phase spectrum. It is pointed out that the quantity C_n is devoid of any phase information and is only a function of the amplitudes of each harmonic contribution in the series. From (1-4) it is seen that C_0 is simply the average value of $g(t)$ over the period T. It is easily seen that C_0 translates the reference amplitude of the signal from one level to another without changing the relative shape of the wave form. In signal theory C_0 is referred to as the dc component of the signal.

It should also be noted that in (1-2a, b) and (1-4) the constants a_n, b_n, or C_n are independent of the time, t, and only a function of the fundamental frequency, ω_0. Thus another way to look at these relations is to consider a_n, b_n, or C_n as the Fourier transform of $g(t)$; that is, the time domain signal has been transformed into the frequency domain. As is shown subsequently it is best to think of the Fourier transform of periodic function in the form given by (1-3) and (1-4).

A quantity extremely useful in measuring the magnitude of a signal is the root-mean-square (rms) value defined as

$$g(t)\big|_{\text{rms}} = \left[\frac{1}{T} \int_0^T g^2(t)\, dt \right]^{1/2} \qquad (1\text{-}6)$$

Using (1-3a) and (1-4) in (1-6) yields

$$g(t)\big|_{\text{rms}} = \left(C_0^2 + 2 \sum_{n=1}^{\infty} |C_n|^2 \right)^{1/2} \qquad (1\text{-}7a)$$

or from (1-5a)

$$g(t)\big|_{\text{rms}} = \left[\frac{a_0^2}{4} + \frac{1}{2} \sum_{n=1}^{\infty} (a_n^2 + b_n^2) \right]^{1/2} \qquad (1\text{-}7b)$$

The square of the rms value is called the *mean-square* value in statistics and more commonly in signal analysis the average power (into a 1-ohm resistor). Thus from (1-6) and (1-7a)

$$P_{av} = \frac{1}{T} \int_0^T g^2(t)\, dt = C_0^{\,2} + 2 \sum_{n=1}^{\infty} |C_n|^2 = \langle g^2(t) \rangle_T \qquad (1\text{-}8)$$

where $\langle \cdots \rangle_T$ denotes the average over time. The relationship 1-8 is called *Parseval's theorem*. The physical significance of this result is that the rms value of a periodic signal $g(t)$ is determined only by the magnitude squared of its Fourier amplitude coefficients and is independent of any phase information. It should be further mentioned that the series expansion of the coefficients $|C_n|^2$ given in (1-8) is in reality an expression for the *power spectrum* of $g(t)$. This becomes obvious if one does not record just the total average power P_{av} but plots the individual contribution to P_{av} of the harmonics comprising $g(t)$. The resulting graph is called the power spectrum of $g(t)$. It should be realized that the power spectra plot is a discontinuous plot consisting of discrete lines of amplitude $|C_n|^2$ at the discrete frequencies $n\omega_0$ ($n = 1, 2, 3, \ldots$).

Before proceeding it is important to note the differences introduced by using C_n given by (1-4) as opposed to a_n and b_n given by (1-2a, b). In the former we see that we have introduced the idea of negative harmonics, C_{-n}. This is, of course, not physically possible; however, we use the form for its mathematical utility. With (1-4) we spread the amplitude spectrum over the range $-n\omega_0$ to $+n\omega_0$ ($n = 0, 1, 2, 3, \ldots$) whereas with (1-2a, b) we only use $n\omega_0$ ($n = 0, 1, 2, \ldots$). The difference in representation is accounted for by a factor of 2. Consider the case when $b_n = 0$. Then, from (1-5a) we see that $C_n = a_n/2$. Thus a power spectrum plot using C_n would depict a two-sided spectrum whose amplitude would be one half of those on a one-sided (a_n) plot. From the square of (1-7a, b) we see that the average power from either form yields the same result.

The above results are illustrated by several examples.

1. Square Wave. The mathematical description of a square wave is

$$g(t) = -A \qquad -\frac{T}{2} < t < -\frac{T}{4}$$

$$g(t) = A \qquad -\frac{T}{4} < t < \frac{T}{4} \qquad (1\text{-}9)$$

$$g(t) = -A \qquad \frac{T}{4} < t < \frac{T}{2}$$

wherein A is the amplitude of the wave and $g(t + T) = g(t)$. This is shown in

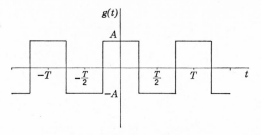

Figure 1-1 Periodic square wave.

Figure 1-1. The symmetric form of the wave was chosen for mathematical convenience. Using (1-9) in (1-2a) and (1-2b) it is an easy matter to show that

$$b_n = a_0 = 0 \tag{1-10}$$

$$a_n = \frac{4A}{n\pi}(-1)^{(n-1)/2} \qquad n = 1, 3, 5, \ldots$$

Hence

$$g(t) = \frac{4A}{\pi} \sum_{n=1,3,5,\ldots}^{\infty} n^{-1}(-1)^{(n-1)/2} \cos n\omega_0 t \tag{1-11}$$

From (1-5a) it is found that

$$|C_n|^2 = \frac{4A^2}{n^2\pi^2} \qquad n = 1, 3, 5, \ldots \tag{1-12}$$

and therefore, from (1-8) the average power is

$$P_{\mathrm{av}} = \frac{8A^2}{\pi^2} \sum_{n=1,3,5,\ldots}^{\infty} n^{-2} = A^2 \tag{1-13}$$

since

$$\sum_{n=1,3,5,\ldots}^{\infty} n^{-2} = \frac{\pi^2}{8} \tag{1-14}$$

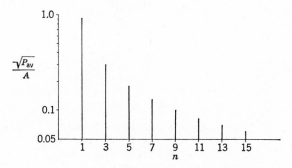

Figure 1-2 Power spectrum of a square wave.

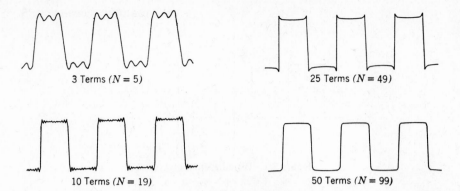

3 Terms *(N = 5)*

25 Terms *(N = 49)*

10 Terms *(N = 19)*

50 Terms *(N = 99)*

Figure 1-3 Square wave as a function of the number of terms in its Fourier series expansion.

$$\cos\left(\frac{2\pi t}{T_0} - \frac{\pi}{12}\right) + \sum_{n=3,5,\ldots}^{101} \frac{(-1)^{(n-1)/2}}{n} \cos\frac{2n\pi t}{T_0}$$

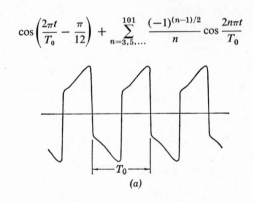

(a)

$$\sum_{n=1,3,\ldots}^{9} \frac{(-1)^{(n-1)/2}}{n} \cos\frac{2n\pi t}{T_0} + 0.75 \sum_{n=11,13,\ldots}^{101} \frac{(-1)^{(n-1)/2}}{n} \cos\frac{2n\pi t}{T_0}$$

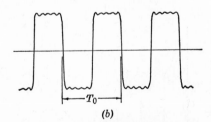

(b)

Figure 1-4 *(a)* Square wave with phase distortion of fundamental frequency. *(b)* Square wave with 25% decrease in the amplitude of the $n \geq 11$ harmonics.

6

Using (1-13) the power spectrum is plotted in Figure 1-2. The log of each of the contributing terms is used to "amplify" the contribution of the higher harmonics.

It is seen from (1-11) that the series is one that converges very slowly. If one were to try to reconstruct (within reasonable accuracy) the square wave with the fewest number of terms it would be found that it would require approximately 50 terms. Thus a periodic square wave requires amplitude and phase information on the order of 100 times the fundamental frequency since the series is summed on odd integers. Figure 1-3 illustrates this.

To emphasize this point further consider two deliberately introduced aberrations to the series given by (1-11). For the first case we give the fundamental-frequency term in (1-11) a phase shift of $-15°$ while the amplitude coefficient remains correct as do all the remaining terms in the series. The result is plotted in Figure 1-4a. For the second case we keep all the phases as they are but introduce an amplitude decrease of 25% to each harmonic contribution after the first five harmonics ($n \geq 11$). The result is plotted in Figure 1-4b. These examples illustrate those often encountered in practice. Hence from the type of distortion that the square wave undergoes one should be able to state qualitatively some of the gross properties of the system through which it is passing.

2. Infinite Pulse Train.   The mathematical description is

$$g(t) = 0 \qquad -\frac{T}{2} < t < -t_0$$

$$g(t) = 2A \qquad -t_0 < t < t_0 \qquad\qquad (1\text{-}15)$$

$$g(t) = 0 \qquad t_0 < t < \frac{T}{2}$$

where $2A$ is the magnitude of the pulse, $2t_0$ is its duration, and $g(t + T) = g(t)$. This is shown in Figure 1-5. Notice from this figure that we should expect

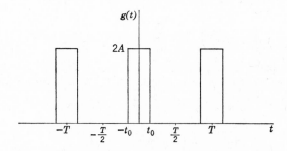

Figure 1-5 Periodic pulse train.

the dc (C_0) term to be nonzero since the entire signal is above the axis. Using either (1-2a, b) or (1-4) it is an easy matter to show that

$$b_n = 0$$

$$a_0 = \frac{8At_0}{T} \tag{1-16}$$

$$a_n = \frac{8At_0}{T} \frac{\sin n\omega_0 t_0}{n\omega_0 t_0} \qquad n = 1, 2, 3, \ldots$$

Hence

$$g(t) = \frac{4At_0}{T}\left(1 + 2\sum_{n=1}^{\infty} \frac{\sin n\omega_0 t_0}{n\omega_0 t_0} \cos n\omega_0 t\right) \tag{1-17}$$

Using (1-5a) yields

$$C_0^{\ 2} = \frac{16A^2 t_0^{\ 2}}{T^2}$$

$$|C_n|^2 = \frac{16A^2 t_0^{\ 2}}{T^2}\left(\frac{\sin n\omega_0 t_0}{n\omega_0 t_0}\right)^2 \qquad n = 1, 2, 3, \ldots \tag{1-18}$$

and, therefore, from (1-8) the average power is

$$P_{\text{av}} = \frac{16A^2 t_0^{\ 2}}{T^2}\left[1 + 2\sum_{n=1}^{\infty}\left(\frac{\sin n\omega_0 t_0}{n\omega_0 t_0}\right)^2\right] \tag{1-19}$$

If $t_0/T = 0.25$, (1-19) becomes

$$P_{\text{av}} = A^2\left(1 + \frac{8}{\pi^2}\sum_{n=1,3,5,\ldots}^{\infty} n^{-2}\right) = 2A^2 \tag{1-20}$$

which is twice the average power in the square wave. In fact, when $t_0/T = 0.25$, if we impress a dc value of magnitude A in (1-9), we obtain (1-15).

3. Triangular Wave Train. The mathematical description is

$$g(t) = \frac{2A}{T}\left(t + \frac{T}{2}\right) \qquad -\frac{T}{2} < t < 0$$

$$g(t) = \frac{2A}{T}\left(\frac{T}{2} - t\right) \qquad 0 < t < \frac{T}{2} \tag{1-21}$$

where T is the period, A is the maximum amplitude, and $g(t) = g(t + T)$. This is shown in Figure 1-6. Again using (1-2a, b) it is found that

$$b_n = 0$$

$$a_0 = A$$

$$a_n = \frac{4A}{n^2 \pi^2} \qquad n = 1, 3, 5, \ldots \tag{1-22}$$

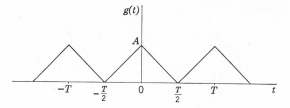

Figure 1-6 Periodic triangular wave.

Hence

$$g(t) = A\left(\frac{1}{2} + \frac{4}{\pi^2} \sum_{n=1,3,5,\dots}^{\infty} n^{-2} \cos n\omega_0 t\right) \tag{1-23}$$

The average power is

$$P_{\mathrm{av}} = A^2\left(\frac{1}{4} + \frac{8}{\pi^4} \sum_{n=1,3,5,\dots}^{\infty} n^{-4}\right) = \frac{A^2}{3} \tag{1-24}$$

since

$$\sum_{n=1,3,5\dots} n^{-4} = \frac{\pi^4}{96}$$

1-3.2 Harmonic Distortion

In many applications the periodic wave form is a single sinusoid of frequency ω_0. However, some systems through which this single frequency sinusoid must pass distort it by introducing harmonics not present in the original signal. It is easily seen from the results of the preceding section that any other periodic wave form except a pure sinusoid will be comprised of an infinite set of weighted harmonics of the fundamental frequency. One is therefore led to the following definition of the total percentage harmonic distortion, %HD:

$$\%\mathrm{HD} = \frac{1}{|C_1|}\left(\sum_{n=2}^{\infty} |C_n|^2\right)^{\frac{1}{2}} \times 100\% \tag{1-25}$$

Notice that except for the omission of the dc value the numerator of (1-25) is simply the difference of the total average power, P_{av}, and the "power" of the fundamental frequency component. For $\%\mathrm{HD} \leqq 10\%$, $|C_1|$ can be replaced by P_{av} to a very good approximation.

To illustrate (1-25) in an extreme case, let us determine the %HD of a square wave. Using (1-12) and (1-14) yields

$$\%\mathrm{HD} = \left(\frac{\pi^2}{8} - 1\right)^{\frac{1}{2}} \times 100\% = 23.3\% \tag{1-26}$$

1-3.3   Correlation

It has already been indicated that one way to measure a signal is to determine its rms value. Another useful and very powerful measure is to determine the degree of "likeness" one signal has with respect to another. Such a measure is discussed below.

Consider two different periodic signals $f_1(t)$ and $f_2(t)$ of the same period T. Form the quantity

$$R_{12}(\tau) = \frac{1}{T} \int_{-T/2}^{T/2} f_1(t) f_2(t + \tau) \, dt \qquad (1\text{-}27)$$

where τ is a continuous time displacement in the range $(-\infty, \infty)$, independent of t. Since $f_1(t)$ and $f_2(t)$ are periodic functions of period T, they can be represented as

$$f_1(t) = \sum_{n=-\infty}^{\infty} C_n e^{jn\omega_0 t} \qquad (1\text{-}28a)$$

and

$$f_2(t) = \sum_{n=-\infty}^{\infty} C'_n e^{jn\omega_0 t} \qquad (1\text{-}28b)$$

where

$$C_n = \frac{1}{T} \int_{-T/2}^{T/2} f_1(t) e^{-jn\omega_0 t} \, dt \qquad (1\text{-}29a)$$

and

$$C'_n = \frac{1}{T} \int_{-T/2}^{T/2} f_2(t) e^{-jn\omega_0 t} \, dt \qquad (1\text{-}29b)$$

Substituting (1-28b) into (1-27) and using (1-29a) yields

$$R_{12}(\tau) = \sum_{n=-\infty}^{\infty} C'_n C_{-n} e^{jn\omega_0 \tau} \qquad (1\text{-}30a)$$

Letting $C'_n C_{-n} = C''_n$, the quantity $R_{12}(\tau)$ is itself expressed in a Fourier series; namely,

$$R_{12}(\tau) = \sum_{n=-\infty}^{\infty} C''_n e^{jn\omega_0 \tau} \qquad (1\text{-}30b)$$

Relation 1-30b implies that

$$C''_n = \frac{1}{T} \int_{-T/2}^{T/2} R_{12}(\tau) e^{-jn\omega_0 \tau} \, d\tau \qquad (1\text{-}31)$$

Thus C''_n can be considered the Fourier transform of $R_{12}(\tau)$. Equations 1-27 and 1-30a give

$$\frac{1}{T} \int_{-T/2}^{T/2} f_1(t) f_2(t + \tau) \, dt = \sum_{n=-\infty}^{\infty} C'_n C_{-n} e^{jn\omega_0 \tau} \qquad (1\text{-}32)$$

Equation 1-32 is the *correlation theorem* for periodic functions. The left-hand side of (1-32) involves a combination of three important operations:

1. One of the periodic functions is given a time displacement τ.
2. The displaced function is multiplied by the other periodic function.
3. The product is averaged by integration over a complete period.

These steps are repeated for every value of τ so that a function is generated. This combination of the three operations, namely, *displacement, multiplication*, and *integration*, is termed *correlation*. It will be shown in the next sections that correlation is by no means limited to periodic functions but can be used with aperiodic and random processes.

The quantity $R_{12}(\tau)$ is called the *cross-correlation* function of $f_1(t)$ and $f_2(t)$. This is the most general case. When $f_1(t) = f_2(t) = f(t)$, and hence $C_n' = C_n$, relations 1-27 and 1-30a give

$$R(\tau) = \sum_{n=-\infty}^{\infty} |C_n|^2 e^{jn\omega_0\tau} = C_0 + 2\sum_{n=1}^{\infty} |C_n|^2 \cos n\omega_0\tau \qquad (1\text{-}33)$$

and inversely, from (1-31),

$$|C_n|^2 = \frac{1}{T}\int_{-T/2}^{T/2} R(\tau)e^{-jn\omega_0\tau}\, d\tau \qquad (1\text{-}34)$$

The quantity $R(\tau)$ is known as the *autocorrelation* function of $f(t)$. It is noted that $R(\tau)$ is a maximum at $\tau = 0$ or any integer multiple of the period, T. It should also be emphasized that the correlation function of two periodic functions is itself a periodic function. This is an extremely useful property in certain applications.

Thus relations 1-33 and 1-34 state that the autocorrelation function and the power spectrum of a periodic function are Fourier transforms of each other. Also, since the power spectrum of a periodic function is independent of the phase angles of its harmonics, it follows that periodic functions having the same harmonic amplitudes but differing in their phase angles have the same autocorrelation function.

In the special case when $\tau = 0$, (1-32) and (1-33) yield

$$R(0) = \frac{1}{T}\int_{-T/2}^{T/2} f^2(t)\, dt = \sum_{n=-\infty}^{\infty} |C_n|^2 = P_{av} \qquad (1\text{-}35)$$

which is Parseval's theorem already obtained in (1-8). This fact is extremely useful when these definitions are extended to aperiodic and random functions. Equation 1-35 is the average power in a 1 ohm resistor.

Consider the following examples of correlation.

1. Autocorrelation of a Symmetric Square Wave. Using the results obtained in (1-12), (1-33) reduces to

$$R(\tau) = \frac{8A^2}{\pi^2} \sum_{n=1,3,5,\dots}^{\infty} n^{-2} \cos n\omega_0\tau \tag{1-36}$$

and

$$R(0) = \frac{8A^2}{\pi^2} \sum_{n=1,3,5,\dots}^{\infty} n^{-2} = A^2 = P_{av} \tag{1-37}$$

It is noticed that (1-36) is the same, except for a dc value, as that given by (1-23) for a triangular wave. Hence the autocorrelation function of a periodic square wave is a periodic triangular wave. The result given by (1-37) confirms that previously obtained in (1-13).

2. Autocorrelation of a Sinusoid. The mathematical expression is

$$f(t) = A_0 \cos \omega_0 t \tag{1-38}$$

where ω_0 is the angular frequency. Substituting (1-38) into (1-2a, b) yields

$$b_n = a_0 = 0$$
$$a_1 = A_0$$
$$a_n = 0 \qquad n \geq 2 \tag{1-39}$$

Hence from (1-5a), $C_0 = 0$, $C_1 = A_0/2$, and $C_n = 0$, $n \geq 2$. Using (1-33) gives

$$R(\tau) = \frac{A_0^2}{2} \cos \omega_0\tau \tag{1-40}$$

Thus the autocorrelation of a sinusoid is another sinusoid of the same period. When $\tau = 0$, (1-40) gives the average power of a sinusoid.

3. Cross-Correlation of a Sinusoid and a Pulse Train. Consider the periodic pulse train given by (1-15) and the sinusoid given by (1-38). Using (1-16), (1-39), and (1-5a) in (1-30a) yields

$$R_{12}(\tau) = 4AA_0 \left(\frac{t_0}{T}\right)\left(\frac{\sin 2\pi t_0/T}{2\pi t_0/T}\right) \cos \omega_0\tau \tag{1-41}$$

As already indicated, when $t_0/T = 0.25$, the pulse train became a square wave as given in (1-9) except for a positive dc value of magnitude A. In this case (1-41) becomes

$$R_{12}(\tau) = \frac{2AA_0}{\pi} \cos \omega_0\tau \tag{1-42}$$

When $t_0/T \ll 1$, the pulse train becomes a series of very narrow pulses and (1-41) becomes

$$R_{12}(\tau) \cong 4AA_0 \frac{t_0}{T} \cos \omega_0 \tau \qquad (1\text{-}43)$$

which is equal to that obtained in (1-40) if $4At_0/T = A_0/2$.

Relations 1-41 to 1-43 show that cross-correlation of a periodic pulse train, square wave, or a train of very narrow pulses with a sinusoid of the same frequency results in a sinusoidal correlation function having the same frequency as the original sinusoid. This property is the basis of many waveform processing techniques. In fact, it can be shown that the cross-correlation of a pulse train with any periodic function results in a periodic correlation function. The exact form of the resulting correlation function depends, of course, on the shape of the periodic function itself. If the pulse train and periodic function are not in phase, the resulting correlation function is still periodic except that it has a maximum at the points in time corresponding to the amount of phase between them.

1-4   Aperiodic Signals

1-4.1   Fourier Transforms

Since the Fourier series described in the preceding section is only valid for periodic signals, new methods must be developed. An aperiodic signal may be considered as one with infinite period. If this view is accepted, then the following approach to an extension of the Fourier series given by (1-3a) and (1-4) can be given. The fundamental frequency of the periodic wave form is $1/T$. In addition, this value is the spacing between adjacent harmonics in the spectrum. Thus $\Delta f = 1/T$ or $\Delta \omega = 2\pi/T = \omega_0$. Then, (1-3a) and (1-4) become

$$f(t) = \sum_{n=-\infty}^{\infty} C_n e^{j(n\Delta\omega)t} \qquad (1\text{-}44a)$$

$$C_n = \Delta f \int_{-\frac{1}{2}\Delta f}^{\frac{1}{2}\Delta f} f(t)\, e^{-j(n\Delta\omega)t}\, dt \qquad (1\text{-}44b)$$

It is seen that as $\Delta f \to 0$, $C_n \to 0$; but the ratio $C_n/\Delta f$ may remain finite. This ratio will in fact remain finite if the definite integral

$$\int_{-\infty}^{\infty} f(t)\, e^{-j(n\Delta\omega)t}\, dt = \text{finite value} \qquad (1\text{-}45)$$

It can be shown that (1-45) is satisfied if

$$\int_{-\infty}^{\infty} |f(t)|\, dt = \text{finite value} \qquad (1\text{-}46)$$

In (1-44b), C_n represents the complex amplitude of the nth harmonic. As Δf becomes smaller, the harmonics in a given frequency band crowd together, so that eventually the spacing between them becomes infinitesimally small (approaches zero). In the Fourier series representation, if ω is the radian frequency of the nth harmonic and the fundamental radian frequency is $\Delta\omega$, then $n\,\Delta\omega = \omega$. The amplitude of this harmonic is C_n. As $\Delta\omega \to 0$, $n \to \infty$, in order that the product $n\,\Delta\omega$ equals a finite ω. In the limit, instead of having discrete harmonics corresponding to n, every value of ω is allowed, and the discrete parameter n is replaced by the continuous variable $\omega = n\,\Delta\omega$. Thus if

$$\lim_{\Delta f \to 0} \frac{C_n}{\Delta f} = F(\omega) \qquad (1\text{-}47)$$

then (1-44b) becomes

$$F(\omega) = \int_{-\infty}^{\infty} f(t)e^{-j\omega t}\,dt \qquad (1\text{-}48)$$

Equation 1-48 is referred to as the complex *Fourier transform* of $f(t)$. It should be noticed that $F(\omega)$ is now a function of *all* frequencies, not just discrete frequencies as given in the periodic counterpart 1-4.

Returning to (1-44a), letting $\Delta f = \Delta\omega/2\pi$, and multiplying by the factor $\Delta f/\Delta f$ gives

$$f(t) = \frac{1}{2\pi} \sum_{n=-\infty}^{\infty} \frac{C_n}{\Delta f} e^{j(n\Delta\omega)t}\,\Delta\omega$$

Taking the limit of the above relation as $\Delta\omega \to 0$ yields the definition of an integral; thus

$$f(t) = \frac{1}{2\pi} \int_{-\infty}^{\infty} F(\omega)e^{j\omega t}\,d\omega \qquad (1\text{-}49)$$

The function $f(t)$ in (1-49) is called the *inverse* Fourier transform of $F(\omega)$. Relations 1-48 and 1-49 comprise the Fourier transform pair.

In general $F(\omega)$ is complex. Thus

$$F(\omega) = R(\omega) - jX(\omega) = A(\omega)e^{-j\theta(\omega)} \qquad (1\text{-}50)$$

where

$$A(\omega) = [R^2(\omega) + X^2(\omega)]^{\frac{1}{2}}$$

and

$$\theta(\omega) = \tan^{-1}\frac{X(\omega)}{R(\omega)}$$

where $A(\omega)$ is the amplitude density spectrum and $\varphi(\omega)$ the phase density spectrum of $F(\omega)$. These are the aperiodic counterparts to (1-5).

The Fourier transform pair has many properties; below are stated without proof several of the important ones.

1. *Convolution*
a. If

$$F(\omega) = \int_{-\infty}^{\infty} f(t)e^{-j\omega t} dt$$

where $f(t) = g(t) \cdot h(t)$, and if

$$G(\omega) = \int_{-\infty}^{\infty} g(t)e^{-j\omega t} dt$$

$$H(\omega) = \int_{-\infty}^{\infty} h(t)e^{-j\omega t} dt$$

then

$$F(\omega) = \frac{1}{2\pi} \int_{-\infty}^{\infty} H(\omega - \alpha)\omega(\alpha)\, d\alpha = \frac{1}{2\pi} \int_{-\infty}^{\infty} H(\alpha)G(\omega - \alpha)\, d\alpha \qquad (1\text{-}51)$$

b. Conversely, if

$$f(t) = \frac{1}{2\pi} \int_{-\infty}^{\infty} F(\omega)e^{j\omega t} d\omega$$

where $F(\omega) = G(\omega) \cdot H(\omega)$, and if

$$g(t) = \frac{1}{2\pi} \int_{-\infty}^{\infty} G(\omega)e^{j\omega t} d\omega$$

$$h(t) = \frac{1}{2\pi} \int_{-\infty}^{\infty} H(\omega)e^{j\omega t} d\omega$$

then

$$f(t) = \int_{-\infty}^{\infty} g(t - \alpha)h(\alpha)\, d\alpha = \int_{-\infty}^{\infty} g(\alpha)h(t - \alpha)\, d\alpha \qquad (1\text{-}52)$$

2. *Multiplication Theorem*

$$\int_{-\infty}^{\infty} f(t)\, g(t)\, dt = \frac{1}{2\pi} \int_{-\infty}^{\infty} F(\omega)\, G^*(\omega)\, d\omega = \frac{1}{2\pi} \int_{-\infty}^{\infty} F^*(\omega)\, G(\omega)\, d\omega \qquad (1\text{-}53)$$

where $[f(t), F(\omega)]$ and $[g(t), G(\omega)]$ constitute Fourier transform pairs given by (1-48) and (1-49), and the asterisk (*) denotes the complex conjugate of the original function.

3. *Time Scaling*
If b is a real constant then $[f(bt), 1/|b|\, F(\omega/b)]$ form a Fourier transform pair. When $b > 1$ the signal $f(t)$ is compressed in time and expanded in the frequency scale.

4. Real-Time Functions

If $f(t)$ is real, then from (1-49) and (1-50) we find

$$R(\omega) = \int_{-\infty}^{\infty} f(t) \cos \omega t \, dt$$

$$X(\omega) = \int_{-\infty}^{\infty} f(t) \sin \omega t \, dt$$

Thus $R(\omega)$ is an even function of ω and $X(\omega)$ is an odd function of ω; that is $R(-\omega) = R(\omega)$ and $X(-\omega) = -X(\omega)$. Therefore $F(-\omega) = F^*(\omega)$. This property is useful when we consider two-sided spectra in Chapter 2.

1-4.2 Spectral Density

The aperiodic counterpart to the mean-square value for a periodic signal follows directly from (1-53). If in (1-53), $f(t) = g(t)$, then

$$\int_{-\infty}^{\infty} f^2(t) \, dt = \frac{1}{2\pi} \int_{-\infty}^{\infty} |F(\omega)|^2 \, d\omega \tag{1-54}$$

since $F(\omega)F^*(\omega) = |F(\omega)|^2$. The quantity $|F(\omega)|^2$ is called the energy density function or the energy spectral density or just the spectral density and is denoted by $S(\omega) = |F(\omega)|^2$. Equation 1-54 is Parseval's theorem for aperiodic functions. Note, however, the sharp contrast between this result and that given for periodic signals in (1-8). In relation 1-8 the *power* was finite, but for aperiodic functions (1-54) shows that the total *energy* is finite [recall (1-46)]. Thus the spectrum of $f(t)$ is an energy density spectrum, which when integrated over all frequencies will yield a value equal to the total energy, E, of the time function, $f(t)$. Thus

$$E = \int_{-\infty}^{\infty} f^2(t) \, dt = \frac{1}{\pi} \int_{0}^{\infty} S(\omega) \, d\omega = 2 \int_{0}^{\infty} S(f) \, df \tag{1-55}$$

since $S(\omega)$ is an even function of ω.

At this point it is worthwhile to examine closely why for periodic signals one talks about power whereas for aperiodic signals one talks about energy. From elementary electronic theory we note that the instantaneous power into a resistive load is $P_i = e(t)i(t)$ where $e(t)$ is the instantaneous voltage and $i(t)$ the instantaneous current. From Ohm's law we have the alternate form $P_i = e^2(t)/R$. If we assume that $R = 1$ ohm, then $P_i = e^2(t)$. The units of P_i are watts. If $e(t)$ is periodic, then the average power P_{av} is given by (1-8) where $e(t)$ replaces $g(t)$. Notice that the value of the integral in (1-8) has the units of energy (watts-second) which when divided by the period T, gives back watts. On the other hand the units of the integral in (1-55) are again energy but since there is not any period over which to average, the units remain those of energy. Hence it is easy to see that $S(\omega)$ has the units watts-second^2 or what is dimensionally equivalent, watts-second per hertz. For this reason $S(\omega)$ is called the energy density, the amount of energy per frequency. The same result can also be seen from (1-47) and the definition of $S(\omega)$.

The above results are illustrated by several examples.

1. Consider a rectangular pulse of duration $2t_0$ and amplitude A. Mathematically this is expresed as

$$f_1(t) = A \qquad -t_0 < t < t_0$$
$$f_1(t) = 0 \qquad |t| > t_0 \tag{1-56}$$

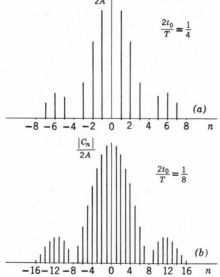

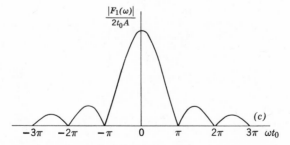

Figure 1-7 Comparison of the amplitude spectrum of a periodic pulse train for several values of $2t_0/T$ to a single pulse of duration $2t_0$.

Then the amplitude spectrum $F_1(\omega)$ is, from (1-48),

$$F_1(\omega) = A \int_{-t_0}^{t_0} e^{-j\omega t} \, dt = 2t_0 A \frac{\sin \omega t_0}{\omega t_0} \tag{1-57}$$

It is seen that the amplitude spectrum for a single pulse is a continuous spectrum whereas its periodic counterpart, the pulse train, contains only discrete frequencies in the transformed plane. It is worthwhile to plot the results obtained for the periodic pulse train in the transformed plane as given by (1-18) for various values of the ratio $2t_0/T$ and compare it to that given by (1-57) for a single pulse. This is shown in Figure 1-7. The total energy of the pulse (dissipated into a 1-ohm resistor) is easily seen to be $E = 2A^2 t_0$. It should be pointed out that the negative part of the rectangular pulse from $-t_0$ to 0 is not physically realizable. It was chosen for mathematical convenience only. If this was not used, the form of $F_1(\omega)$ would be altered but the spectral energy $S(\omega)$ would remain the same.

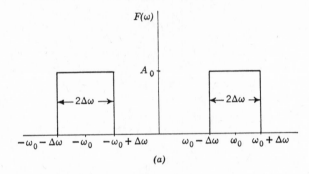

(a)

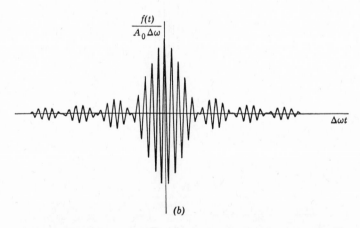

(b)

Figure 1-8   (a) Rectangularly symmetrical amplitude spectrum. (b) Time-domain function of a.

From Figure 1-7c we see that the majority of the frequency information is contained between $-\pi \leqq \omega t_0 \leqq \pi$. Considering the physically realizable spectrum ($\omega \geqq 0$)—recall the discussion preceding (1-9) which also applies to the Fourier transform—we see that one could expect the time-varying rectangular pulse to be reasonably reproduced with the frequencies $0 \leqq f \leqq 1/2t_0$. Thus (1-57) provides us with the very useful conclusion that most of the frequency information contained in a pulse is in a frequency band proportional to the reciprocal of its duration.

2. Now consider the inverse case. Let the square root of the frequency spectrum, that is, the amplitude spectrum be

$$F(\omega) = A_0 \qquad \omega_0 - \Delta\omega < \omega < \omega_0 + \Delta\omega; \; -\omega_0 - \Delta\omega < \omega < -\omega_0 + \Delta\omega$$
$$\quad\;\; = 0 \qquad \text{elsewhere}$$

This is shown in Figure 1-8a. Thus from (1-49), we obtain

$$f(t) = \frac{2A_0}{\pi} \Delta\omega \left(\frac{\sin \Delta\omega t}{\Delta\omega t}\right) \cos \omega_0 t \tag{1-58}$$

We shall use this result later in this chapter and again in Chapter 2 where its significance will become more apparent. It is plotted in Figure 1-8b.

1-4.3 Dirac Delta Function

Let us now go back and examine the result given in (1-57) which is plotted in Figure 1-7c. Let $\tau = 2t_0$ and assume that as $\tau \to 0$, $A \to \infty$, such that the product $A\tau \to 1$. Then the spectrum shown in Figure 1-7c would tend to that shown in Figure 1-9. More important, however, is the peculiar nature of the time varying wave form—a spike of infinite amplitude and zero width. Such a function is mathematical nonsense. However, it does lead to the definition of the Dirac delta function given as

$$\int_{-\infty}^{\infty} f(t)\, \delta(t - t_0)\, dt = f(t_0) \tag{1-59}$$

where $\delta(t - t_0)$ is the Dirac delta function. Notice that in the definition only

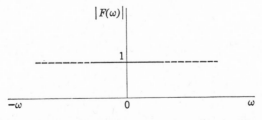

Figure 1-9 Amplitude spectrum of a spike in the time domain.

the integral of the delta function is defined. Also, if $f(t) = 1$ and $t_0 = 0$, (1-59) gives

$$\int_{-\infty}^{\infty} \delta(t)\, dt = 1 \qquad (1\text{-}60)$$

Thus $\delta(t)$ is sometimes thought of as a function with the properties

$$\delta(t) = 0 \qquad t \neq 0$$
$$\delta(t) = \infty \qquad t = 0$$

In subsequent use of the delta function it will appear on occasion without the integral sign; however, this does not mean that the integral will not be taken; it only means that integration will eventually be performed. The delta function is called a symbolic function since it is not a function in the real sense. A rigorous justification of it is given in the theory of distributions.

The Fourier transform of the delta function yields

$$F(\omega) = \int_{-\infty}^{\infty} \delta(t - t_0)e^{-j\omega t}\, dt = e^{-j\omega t_0} \qquad (1\text{-}61)$$

This spectrum is also given by that shown in Figure 1-9. Now consider what the time function is if $F(\omega) = \delta(\omega - \omega_0)$. Hence from (1-49)

$$f(t) = \frac{1}{2\pi} \int_{-\infty}^{\infty} \delta(\omega)e^{j\omega t}\, d\omega = \frac{1}{2\pi} e^{j\omega_0 t} \qquad (1\text{-}62)$$

Using the definition of the Fourier transform pair, (1-61) and (1-62) yields the interesting result

$$\delta(t) = \frac{1}{2\pi} \int_{-\infty}^{\infty} e^{j\omega t}\, d\omega \qquad (1\text{-}63a)$$

$$\delta(\omega) = \frac{1}{2\pi} \int_{-\infty}^{\infty} e^{-j\omega t}\, dt \qquad (1\text{-}63b)$$

Now consider a time domain function of the form

$$f(t) = \tfrac{1}{2}\delta(t - t_0) + \tfrac{1}{2}\delta(t + t_0)$$

Then, from (1-61),

$$F(\omega) = \tfrac{1}{2}e^{-j\omega t_0} + \tfrac{1}{2}e^{j\omega t_0} = \cos \omega t_0$$

Conversely, if

$$F(\omega) = \pi\delta(\omega - \omega_0) + \pi\delta(\omega + \omega_0)$$

then,

$$f(t) = \tfrac{1}{2}e^{j\omega_0 t} + \tfrac{1}{2}e^{-j\omega_0 t} = \cos \omega_0 t \qquad (1\text{-}64)$$

This result, (1-64), is of fundamental importance because it provides a means of dealing with discrete frequencies (generated by periodic functions) at the same time as with continuous spectra (generated by aperiodic functions). A similar procedure yields results for $\sin \omega t_0$ and $\sin \omega_0 t$.

1-4.4   Correlation

In a manner analogous to the case of the periodic signals, the definition for the correlation function of an aperiodic function can be given. Using the multiplication theorem with $g(t) = f_2(t + \tau)$ and noticing that

$$e^{-j\omega\tau}F_2(\omega) = \int_{-\infty}^{\infty} f_2(t + \tau)e^{-j\omega\tau}\, dt$$

gives

$$\int_{-\infty}^{\infty} f_1(t)f_2(t + \tau)\, dt = \frac{1}{2\pi}\int_{-\infty}^{\infty} F_1(\omega)F_2^*(\omega)e^{j\omega\tau}\, d\omega \qquad (1\text{-}65)$$

where $[f_1(t), F_1(\omega)]$ and $[f_2(t), F_2(\omega)]$ constitute Fourier transform pairs. The quantity $F_1(\omega)F_2^*(\omega)$ is denoted by

$$S_{12}(\omega) = F_1(\omega)F_2^*(\omega) \qquad (1\text{-}66)$$

where $S_{12}(\omega)$ is called the cross-spectral density spectrum. The quantity on the left-hand side of (1-65) is called the cross-correlation of $f_1(t)$ and $f_2(t)$ and is denoted as before by $R_{12}(\tau)$. Thus

$$R_{12}(\tau) = \int_{-\infty}^{\infty} f_1(t)f_2(t + \tau)\, dt = \frac{1}{2\pi}\int_{-\infty}^{\infty} S_{12}(\omega)e^{j\omega\tau}\, d\omega \qquad (1\text{-}67)$$

It is seen that as in the analogous case of the periodic signals, the cross-correlation function $R_{12}(\tau)$ and the energy density function $S_{12}(\omega)$ form a Fourier transform pair:

$$R_{12}(\tau) = \frac{1}{2\pi}\int_{-\infty}^{\infty} S_{12}(\omega)e^{j\omega\tau}\, d\omega \qquad (1\text{-}68a)$$

and

$$S_{12}(\omega) = \int_{-\infty}^{\infty} R_{12}(\tau)e^{-j\omega\tau}\, d\tau \qquad (1\text{-}68b)$$

Equation 1-68a is used when time-delay information is desired and (1-68b) is used when frequency information is desired.

When $f_1(t) = f_2(t) = f(t)$, the above results reduce to the definition of the autocorrelation function. Thus

$$R(\tau) = \int_{-\infty}^{\infty} f(t)\, f(t + \tau)\, dt = \frac{1}{2\pi}\int_{-\infty}^{\infty} S(\omega)e^{j\omega\tau}\, d\omega \qquad (1\text{-}69)$$

where $S(\omega) = |F(\omega)|^2$. Equation 1-68a, b gives

$$R(\tau) = \frac{1}{2\pi}\int_{-\infty}^{\infty} S(\omega)e^{j\omega\tau}\, d\omega \qquad (1\text{-}70a)$$

$$S(\omega) = \int_{-\infty}^{\infty} R(\tau)e^{-j\omega\tau}\, d\tau \qquad (1\text{-}70b)$$

It is seen from (1-69) that when $\tau = 0$, (1-54) is obtained. Thus the auto-correlation function of an aperiodic signal at $\tau = 0$ is equal to the total energy in the signal.

To illustrate these results let us consider several examples that prove useful in subsequent chapters.

1. Consider the case where $S(\omega) = N = $ constant. Using (1-70a) and (1-63a) yields

$$R(\tau) = N\delta(\tau) \tag{1-71}$$

2. Let

$$S(\omega) = \frac{\beta^2 K}{\omega^2 + \beta^2} \tag{1-72}$$

Then from (1-70a) we obtain

$$R(\tau) = \frac{\pi K \beta}{2} e^{-\beta|\tau|} \tag{1-73}$$

3. Let

$$S(\omega) = 4Kb^2a^2[(a^2 - \omega^2)^2 + 4b^2\omega^2]^{-1} \tag{1-74}$$

Then from (1-70a) we obtain

$$R(\tau) = \pi K b e^{-b|\tau|}\left(\cos c\tau + \frac{b}{c}\sin c|\tau|\right) \tag{1-75}$$

where $c^2 = a^2 - b^2$. In the case when $b/c \ll 1$, (1-75) reduces to

$$R(\tau) \approx \pi K b e^{-b|\tau|}\cos c\tau \tag{1-76}$$

1-4.5 Aperiodic Functions Made Periodic

It is sometimes necessary and often extremely useful to make an aperiodic signal, say $f(t)$, periodic in the interval T. (This can be done by "capturing" the aperiodic signal and then repeating it in its entirety every period T.) Hence if $f(t)$ is zero outside the interval $-T_0/2 < t < T_0/2$, (1-48) gives

$$F(\omega) = \int_{-T_0/2}^{T_0/2} f(t)e^{-j\omega t}\,dt \tag{1-77}$$

If $f(t)$ is now made periodic on some interval $T(T > T_0)$, (1-4) gives

$$C_n = \frac{1}{T}\int_{-T/2}^{T/2} f(t)e^{-jn\omega_0 t}\,dt \tag{1-78}$$

where $\omega_0 = 2\pi/T$. Recalling the heuristic argument given in the derivation of (1-48) it is desirable, in order for (1-78) to approximate closely (1-77), for T to be as large as possible (the exact limits depending on the signal processing technique being used—see Chapters 2 and 5). Referring to Figure 1-7 it is seen that if $T_0/T \ll 1$, the envelope of the amplitude of the frequency distribution will closely approximate the continuous spectrum given by (1-77). This approximation is the basis for using tape loops and digital-recirculating delay lines which are discussed in Chapters 5 and 7, respectively.

1-5   Random Signals

1-5.1   Probability Density Functions

It has been noted that both periodic and aperiodic functions of time are completely determined in the time domain for every value of the independent variable and can be specified over the entire range of frequencies by a transformation from the time domain to the frequency domain. Random signals, however, are characterized by not being subject to precise prediction. When a random wave form is examined it reveals that its average amplitude appears to be in the *vicinity* of some amplitude more often than others. The word "vicinity" must be used for it is impossible to specify the amount of time spent *at* a given amplitude. When a signal takes on a specified value of the amplitide, the intercept of the wave form with a specified amplitude value defines only a point. The projection of this point on a time axis defines, therefore, just a point in time and not a time interval.

Consider a random process in which the range of variation of the random variable is continuous. In this instance the variable is called a *continuous random variable*. It should be noted that the term *continuous* refers to the range of variation *not* the values of the variable in a sequence of trials. In such a sequence the random variable takes values in this range according to chance, and it is clear that the sequence of values cannot be continuous although the range of possible values is continuous. To characterize a random variable whose range of variation is continuous, a probability function with a continuous argument representing the range becomes necessary. In this function, an element of the range is dx, and with it a probability is associated. Let $P_\xi(x)$ be the function in question.

To determine $P_\xi(x)$ we note that the incremental probability, $\Delta \mathscr{P}(x)$, of ξ, being in the interval $(x, x + \Delta x)$, is

$$\Delta \mathscr{P}(x) = \mathscr{P}(x + \Delta x) - \mathscr{P}(x) \tag{1-79}$$

If $\Delta x \to 0$, $\Delta \mathscr{P}(x)$ would also go to zero and no useful information would be obtained. However, if we divide (1-79) by Δx and take the limit as $\Delta x \to 0$, we have

$$P_\xi(x) = \frac{d \mathscr{P}(x)}{dx} = \lim_{\Delta x \to 0} \frac{\mathscr{P}(x + \Delta x) - \mathscr{P}(x)}{\Delta x} \tag{1-80}$$

We call the function $P_\xi(x)$ the probability density function. It is always a real-valued nonnegative function. The product of $P_\xi(x)\, dx$ is an element of area

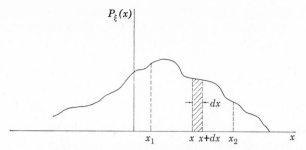

Figure 1-10   A probability density function.

that represents an element of probability. This probability is for the event that, in a single trial, ξ falls in the infinitesimal interval $(x, x + dx)$. Since the range is divided into elements each of length dx and the element of probability $P_\xi(x)\, dx$ is the probability of falling in one of these elements, the density $P_\xi(x)$ is actually a function showing the relative magnitudes of probability for ξ to fall in an element dx at x. For the probability of the event that ξ falls in the infinitesimal interval $(x, x + dx)$, the symbolic form is written

$$\mathscr{P}(x < \xi < x + dx) = P_\xi(x)\, dx$$

Figure 1-10 shows a probability density function. Notice that although only positive x has been considered so far, the results are equally valid for negative x.

If one were to consider the compound event that ξ falls in the interval (x_1, x_2), its probability is expressed as

$$\mathscr{P}(x_1 < \xi < x_2) = \int_{x_1}^{x_2} P_\xi(x)\, dx \tag{1-81}$$

For continuous random variables the total probability is normalized such that

$$\int_{-\infty}^{\infty} P_\xi(x)\, dx = 1 \tag{1-82}$$

To apply the above to time-varying wave forms consider a continuous signal $y(t)$ shown in Figure 1-11. The probability that the amplitude ξ of the given signal will be found between x and $x + dx$ is given by

$$\Delta\mathscr{P}(x) = \mathscr{P}(x < \xi < x + dx) = \lim_{T \to \infty} \left(\frac{\sum_i \Delta t_i}{T} \right) \tag{1-83}$$

where $\sum_i \Delta t_i$ is the total time spent by $y(t)$ within the interval x and $x + dx$ during the time T. Then the probability density function shows the relative

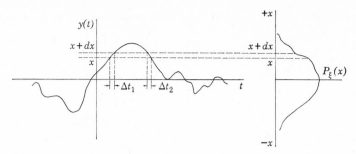

Figure 1-11 Determination of the probability density function for a random wave form.

frequency of occurrence of the various instantaneous values of $y(t)$ in $(x, x + dx)$ and is given by [see (1-80)]

$$P_\xi(x) = \lim_{\substack{dx \to 0 \\ T \to \infty}} \left(\frac{\sum_i \Delta t_i}{T \, \Delta x} \right) \tag{1-84}$$

These results can be extended to two random variables $y(t)$ and $g(t)$. In this case, analogous to (1-83), the joint probability of two random variables is given by

$$\mathscr{P}(u < \xi < u + du, v < \eta < v + dv) = \lim_{T \to \infty} \left(\frac{\sum_i \Delta t_i}{T} \right)$$

where $\sum_i \Delta t_i$ now represents the total time during the time T spent by $y(t)$ and $g(t)$ *simultaneously* in the band $(u, u + du)$ and $(v, v + dv)$. Then the joint probability density function is given by

$$P_{\xi\eta}(u, v) = \lim_{\substack{\Delta u \to 0 \\ \Delta v \to 0 \\ T \to \infty}} \left(\frac{\sum_i \Delta t_i}{T \, \Delta u \, \Delta v} \right)$$

Using reasoning similar to that used to obtain (1-81) yields

$$\mathscr{P}(a_1 < \xi < b_1, a_2 < \eta < b_2) = \int_{a_1}^{b_1} \int_{a_2}^{b_2} P_{\xi\eta}(u, v) \, du \, dv \tag{1-85}$$

The total joint probability density function is normalized, as was (1-82), such that

$$\int_{-\infty}^{\infty} \int_{-\infty}^{\infty} P_{\xi\eta}(u, v) \, du \, dv = 1 \tag{1-86}$$

It is worthwhile to note that when $y(t)$ and $g(t)$ are statistically independent

$$P_{\xi\eta}(u, v) \, du \, dv = P_\xi(u)P_\eta(v) \, du \, dv \qquad (1\text{-}87)$$

and (1-85) becomes

$$\mathscr{P}\,(a_1 < \xi < b_1; a_2 < \eta < b_2) = \int_{a_1}^{b_1} P_\xi(u) \, du \int_{a_2}^{b_2} P_\eta(v) \, dv$$
$$= \mathscr{P}(a_1 < \xi < b_1)\mathscr{P}(a_2 < \eta < b_2) \qquad (1\text{-}88)$$

Equation 1-88 states that the probability of the product of statistically independent random variables is equal to the product of the probability of each random variable.

The results given by (1-81) and (1-88) can be generalized, respectively, to include an arbitrary weight function. Thus

$$\langle g(t, \xi)\rangle = \int_{-\infty}^{\infty} g(t, x)P_\xi(x) \, dx \qquad (1\text{-}89a)$$

$$\langle h(t, \xi, \eta)\rangle = \int_{-\infty}^{\infty} \int_{-\infty}^{\infty} h(t, u, v)P_{\xi\eta}(u, v) \, du \, dv \qquad (1\text{-}89b)$$

where t is a parameter and the symbol $\langle \cdots \rangle$ denotes the expected value of argument; in the above case, $g(t, \xi)$ and $h(t, \xi, \eta)$, respectively.

We now consider several very important special cases of (1-89a, b). Let $g(t, x) = x^m$. Then (1-89a) becomes

$$\langle \xi^m \rangle = \int_{-\infty}^{\infty} x^m P_\xi(x) \, dx \qquad (1\text{-}90)$$

When $m = 1$, $\langle \xi \rangle$ is called the average or mean value of x. When $m = 2$, $\langle \xi^2 \rangle$ is called the mean square of the random variable x. Using these two definitions we obtain another very important description of random processes, the variance, σ_ξ^2, defined as

$$\sigma_\xi^2 = \langle (\xi - \langle \xi \rangle)^2 \rangle = \int_{-\infty}^{\infty} (x - \langle \xi \rangle)^2 P_\xi(x) \, dx$$
$$\sigma_\xi^2 = \langle \xi^2 \rangle - \langle \xi \rangle^2 \qquad (1\text{-}91)$$

Thus it is seen that the variance of ξ is the mean square of ξ minus the square of the mean of ξ. If the mean is zero, $\langle \xi \rangle = 0$, and the variance is equal to the mean-square value. The square root of the variance is the standard deviation of the random variable ξ which is a measure of the amount of dispersion around the mean.

Similar results can be obtained for (1-89b). In particular we record

$$\sigma_{\xi\eta}^2 = \langle (\xi - \langle \xi \rangle)(\eta - \langle \eta \rangle)\rangle = \int_{-\infty}^{\infty} \int_{-\infty}^{\infty} (u - \langle \xi \rangle)(v - \langle \eta \rangle)P_{\xi\eta}(u, v) \, du \, dv$$
$$\sigma_{\xi\eta}^2 = \langle \xi\eta \rangle - \langle \xi \rangle\langle \eta \rangle \qquad (1\text{-}92)$$

In the particular case wherein ξ and η are statistically independent we have from (1-88) that $\langle \xi\eta \rangle = \langle \xi \rangle \langle \eta \rangle$ and, therefore, $\sigma_{\xi\eta}^2 = 0$. This then becomes the definition of linearly independent or uncorrelated random variables.

It is convenient to introduce the standardized variable z corresponding to the random variable x which is defined as

$$z = \frac{x - \mu}{\sigma_x} \tag{1-93}$$

where $\mu = \langle x \rangle$ is the mean and $\sigma_x = \langle (x - \mu)^2 \rangle^{1/2}$ is the standard deviation. Then it is not difficult to show that $\langle z \rangle = 0$ and $\langle z^2 \rangle = 1$. Using this definition and its properties we record for future use several probability density functions.

1. Gaussian

$$P_\xi(z) = \frac{1}{\sqrt{2\pi}} e^{-z^2/2} \tag{1-94}$$

where z is given by (1-93). This type of process is quite prevalent in many physical problems.

2. Chi-Square

$$P_\xi(\chi^2) = \left[2^{n/2}\Gamma\left(\frac{n}{2}\right) \right]^{-1} (\chi^2)^{(n-2)/2} e^{-\chi^2/2} \tag{1-95}$$

where $\Gamma(n/2)$ is the Gamma function and the symbol χ^2 is the chi-square variable. The integer n is the number of degrees of freedom or the number of independent random variables considered. We discuss this distribution in more detail in Section 1-5.3.

3. Rayleigh

$$P_\xi(z) = ze^{-z^2/2} \tag{1-96}$$

This distribution function is the limiting function for the peak values of a narrow-band Gaussian random signal as the bandwidth approaches zero. These three density functions are plotted in Figure 1-12.

1-5.2   Characteristic Functions

A specific form of (1-89a), which is of considerable importance, is obtained by letting $g(t, x) = e^{jxt}$. Then (1-89a) gives

$$\langle e^{jxt} \rangle = p_\xi(t) = \int_{-\infty}^{\infty} e^{jxt} P_\xi(x) \, dx \tag{1-97}$$

This average, denoted by $p_\xi(t)$ (t real), is called the characteristic function of the probability density $P_\xi(x)$. In the terminology of harmonic analysis the characteristic function is the inverse Fourier transform of the probability

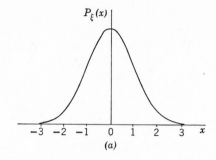

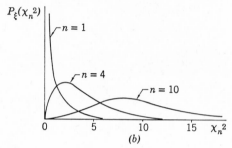

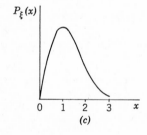

Figure 1-12   The (a) Gaussian, (b) chi-square, and (c) Rayleigh probability density functions.

density function. It is a special transform in that the function under transformation $[P_\xi(x)]$ is real and positive and has an integral over all x equal to 1. To express $P_\xi(x)$ in terms of $p_\xi(x)$ we use (1-49) to obtain the Fourier transform

$$P_\xi(x) = \frac{1}{2\pi} \int_{-\infty}^{\infty} p_\xi(t) e^{-jxt}\, dt \qquad (1\text{-}98)$$

As we shall show next, it is sometimes easier to determine the characteristic function first and then transform it to obtain the probability density than it is to determine the probability density function directly.

From the definition of $p_\xi(t)$ given by (1-97) we see that it is the expected value of $e^{j\xi t}$ where ξ is the amplitude and t is a parameter. According to Cramér (ref. 5, p. 185) the characteristic function of any function $g(\xi)$ is the expected value of $e^{jtg(\xi)}$. Using this fact we can develop an expression to determine the characteristic function and hence the probability density function of a periodic signal. The mean value according to Cramér is then

$$p_\xi(u) = \frac{1}{T}\int_0^T e^{juf(t)}\, dt \tag{1-99}$$

where $f(t)$ is the periodic signal of period T. Substituting (1-99) into (1-98) gives

$$P_\xi(x) = \frac{1}{2\pi}\int_{-\infty}^{\infty}\left[\frac{1}{T}\int_0^T e^{juf(t)}\, dt\right]e^{-jux}\, du \tag{1-100}$$

This result is now used to calculate the probability density functions of several common periodic wave forms.

1. Sine Wave. Let $f(t) = A\sin\omega_0 t$, $\omega_0 = 2\pi/T$. Then

$$P_\xi(x) = \frac{1}{2\pi}\int_{-\infty}^{\infty}\left(\frac{1}{T}\int_0^T e^{juA\sin\omega_0 t}\, dt\right)e^{-jux}\, du$$

$$\begin{aligned}P_\xi(x) &= \pi^{-1}(A^2 - x^2)^{-\frac{1}{2}} &|x| \leq A\\ &= 0 &|x| > A\end{aligned} \tag{1-101}$$

Notice that the $P_\xi(x)$ is independent of the frequency and phase of the sine wave. It is an easy matter to show that (1-82) is satisfied.

2. Square Wave. In this case

$$f(t) = A \qquad 0 < t < \frac{T}{2}$$

$$= -A \qquad \frac{T}{2} < t < T$$

Then

$$P_\xi(x) = \frac{1}{2\pi}\int_{-\infty}^{\infty}\left(\frac{1}{T}\int_0^{T/2} e^{juA}\, dt + \frac{1}{T}\int_{T/2}^T e^{-juA}\, dt\right)e^{-jux}\, du$$

$$P_\xi(x) = \tfrac{1}{2}[\delta(x - A) + \delta(x + A)]$$

wherein (1-63) has been used. It is easily verified that (1-82) is again satisfied.

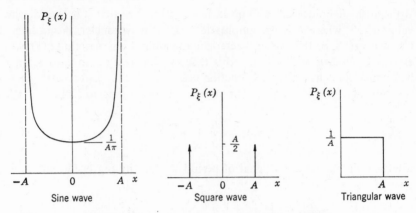

Figure 1-13   Probability density functions of some common periodic wave forms.

3. Triangular Wave.   In this case we use (1-21) to obtain

$$P_\xi(x) = \frac{1}{2\pi} \int_{-\infty}^{\infty} \left(\frac{2e^{jAu}}{T} \int_0^{T/2} e^{-j2Au/T} \, dt \right) e^{-jux} \, du$$

$$P_\xi(x) = A^{-1} \qquad 0 \leqq x \leqq A$$

$$\qquad\quad = 0 \qquad\quad x > A$$

In this particular case (1-84) would have been easier to use.

These results are shown in Figure 1-13.

Application 1-1   Probability Density Function and the Distortion of Periodic Signals

Although the probability density function contains no information about the phase and frequency of a periodic signal, it does contain significant amplitude information in the form of relative frequency of occurrence of amplitude level. One way to look at distortion is to consider it a mechanism which causes a shift in the relative amplitudes of the original signal. This seems a reasonable interpretation if one considers the definition of harmonic distortion given by (1-25). Thus if one knows the probability density function of the undistorted signal (either experimentally or theoretically) any deviation from it indicates the presence of amplitude distortion. The exact percentage of distortion can be determined by comparing the distorted to the undistorted probability density function. More often, however, the detection of distortion from the probability density function is used as a qualitative indicator that distortion is present. This technique is not necessarily restricted to periodic signals. It is also suitable for a random process with a known probability density function.

1-5.3 Chi-Square Revisited

Let us re-examine the chi-square distribution and with it introduce several concepts that will be extremely useful when we consider in the subsequent chapters such topics as bandwidth, averaging times, and resolution. Let $z_i (i = 1, 2, \ldots)$ be a number of random variables each having a Gaussian probability distribution. Furthermore, let $\langle z_i \rangle = 0$ and $\langle z_i^2 \rangle = 1$. Thus the definition of each z_i is given by (1-93). Finally we let z_i be statistically independent; that is, $\langle z_i z_k \rangle = 0$, $i \neq k$ [recall (1-92)ff].

Now consider the sum of the squares of n of these random variables (called degrees of freedom)

$$\chi_n^2 = \sum_{k=1}^n z_k^2 \tag{1-102}$$

The resulting random variable, χ_n^2, is called chi-square. Notice that $\chi_n^2 > 0$ always. It can be shown that

$$\mu = \langle \chi_n^2 \rangle = n \tag{1-103a}$$

$$\sigma^2 = \langle (\chi_n^2 - \langle \chi_n^2 \rangle)^2 \rangle = 2n \tag{1-103b}$$

where μ is the mean and σ^2 is the variance. For purposes here we see that chi-square is a simple random variable that depends only on the parameter n. A measure of the relative variation about the mean, μ, is indicated by

$$\left(\frac{\sigma}{\mu} \right)_{\chi_n^2} = \sqrt{\frac{2}{n}} \tag{1-104}$$

where σ is the standard deviation. As the number of degrees of freedom increases, the actual deviation (dispersion) $\sigma = \sqrt{2n}$ increases, but the relative deviation in comparison with the mean decreases. This behavior is shown in Figure 1-14. Recall from (1-103a) that the mean increases linearly with n, but as just stated the dispersion around the mean decreases. In Figure 1-14 exact tables for the χ_n^2 as given by the integral of (1-95), that is, (1-81) were used to plot limits for 80% of the distribution such that 10% of the samples will lie above the upper limit and 10% below the lower limit.

The chi-square model serves as a useful mode for many more complicated random phenomena resulting from the superposition of a number of essentially additive sources. Unfortunately, in many instances these sources do not have equal weight, and they are not statistically independent. To circumvent this difficulty, (1-104) is modified by evaluating for the phenomenon in question μ and σ and stating that

$$\left(\frac{\sigma}{\mu} \right)_{\text{phenomenon}} = \sqrt{\frac{2}{n_e}} \tag{1-105}$$

where n_e is the equivalent number of statistical degrees of freedom.

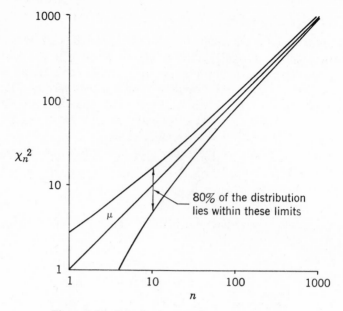

Figure 1-14   Distribution of $\chi_n{}^2$ as a function of n.

It should be noted that the amount of variability obtained from chi-square is very large when n is small. To emphasize the point the data from Figure 1-14 are replotted in Figure 1-15 by scaling down the mean value of every distribution to unity. Thus when $n = 10$, $\chi_n{}^2/n$ lies between 0.49 and 1.60. In other words, 80% of the time the individual samples will lie within -51 and $+60\%$ of the mean. On the other hand, if $n = 500$, we can say that 80% of the time the individual samples will lie within $\pm 8\%$ of the mean.

Let us now consider the case wherein it is known that the distribution is chi-square but the mean μ is unknown. We introduce an associated random variable y such that $y = \mu z/n$, wherein μ is a constant and z is a random variable having a chi-square distribution. It is easily shown that

$$\langle y \rangle = \mu$$

$$\langle (y - \langle y \rangle)^2 \rangle = \frac{2\mu^2}{n}$$

(1-106)

so that the ratio $\sigma/\mu = \sqrt{2/n}$ equals (1-104), the ratio for the chi-square itself. If μ is known, then Figure 1-15 can be used to define the central 80% of the distribution of possible samples; for example, if $n = 50$ the ratio y/μ will lie between 0.75 and 1.26 in 80% of the samples.

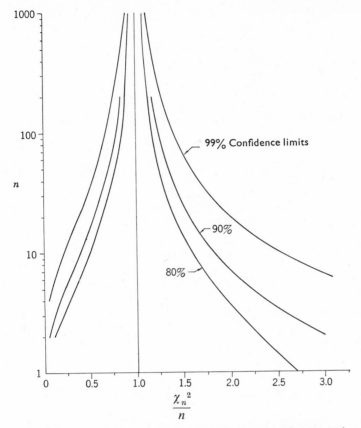

Figure 1-15 Number of n (degrees of freedom) required to have central 80, 90, and 99% of the distribution χ_n^2/n within the indicated limits.

When μ is unknown, we can attempt to measure it by choosing a single sample y_0. Now we use Figure 1-15 to state the *confidence* limits on μ. Considering the previous example for $n = 50$, we see that $y_0/\mu_1 = 1.26$ and $y_0/\mu_2 = 0.75$, and that in any case in which the true mean lies in the band between $\mu_1 = 0.79y_0$ and $\mu_2 = 1.33y_0$, a sample of magnitude y_0 would lie within the central 80% of the true distribution about the true mean. These limits μ_1 and μ_2 are called the confidence limits for the mean.

1-5.4 Nonstationary, Stationary, and Ergodic Processes

We define an individual time history representing a random phenomenon as a *sample* function, denoted as $^kx(t)$. The entire collection of random records or the *ensemble* of the sample functions is called the random process and denoted $\{^kx(t)\}$. For a given value of time $t = t_1$ the average over the ensemble, denoted by $\langle \cdots \rangle_e$, is given by

$$\bar{\mu}_x(t_1) = \langle {}^k x(t_1) \rangle_e = \lim_{n \to \infty} \frac{1}{n} \sum_{k=1}^{n} {}^k x(t_1) \qquad (1\text{-}107)$$

In general, $\bar{\mu}_x(t_1) \neq \bar{\mu}_x(t_2)$, $t_1 \neq t_2$.

Consider now another random process $\{{}^k y(t)\}$. We define the covariance as

$$\rho_{xy}(t_1, t_2) = \langle [{}^k x(t_1) - \bar{\mu}_x(t_1)][{}^k y(t_2) - \bar{\mu}_y(t_2)] \rangle_e \qquad (1\text{-}108)$$

which upon expanding and using (1-107) gives

$$\rho_{xy}(t_1, t_2) = R_{xy}(t_1, t_2) - \bar{\mu}_x(t_1)\bar{\mu}_y(t_2) \qquad (1\text{-}109)$$

where

$$R_{xy}(t_1, t_2) = \langle {}^k x(t_1){}^k y(t_2) \rangle_e = \lim_{n \to \infty} \frac{1}{n} \sum_{k=1}^{n} {}^k x(t_1){}^k y(t_2) \qquad (1\text{-}110)$$

and $R_{xy}(t_1, t_2)$ is the cross-correlation function. We see that the covariance function $\rho_{xy}(t_1, t_2)$ equals $R_{xy}(t_1, t_2)$ when the mean of one of the random processes is zero. It should be noted that

$$R_{xy}(t_1, t_2) = R_{yx}(t_2, t_1)$$

For the case where ${}^k x(t) = {}^k y(t)$ we obtain the definition of the auto-covariance function

$$\rho_{xx}(t_1, t_2) = R_{xx}(t_1, t_2) - \bar{\mu}_x(t_1)\bar{\mu}_x(t_2) \qquad (1\text{-}111)$$

where $R_{xx}(t_1, t_2)$ is the autocorrelation function. A similar expression exists for $\rho_{yy}(t_1, t_2)$.

To investigate how the mean and covariance depend on the time t_1 and t_2 we let $t_1 = t$ and $t_2 = t + \tau$. Then (1-108) and (1-109) become, respectively,

$$\rho_{xy}(t, \tau) = \langle [{}^k x(t) - \bar{\mu}_x(t)][{}^k y(t + \tau) - \bar{\mu}_y(t + \tau)] \rangle_e \qquad (1\text{-}112a)$$

and

$$\rho_{xy}(t, \tau) = R_{xy}(t, \tau) - \bar{\mu}_x(t)\bar{\mu}_y(t + \tau) \qquad (1\text{-}112b)$$

Equations 1-112a, b can be reduced in an obvious manner to the auto-covariance case. In the special case where $\tau = 0$ ($t_1 = t_2$), (1-112b) yields

$$\rho_{xy}(t, 0) = R_{xy}(t, 0) - \bar{\mu}_x(t)\bar{\mu}_y(t) = \sigma_{xy}^2(t) \qquad (1\text{-}113a)$$

and when ${}^k x(t) = {}^k y(t)$

$$\rho_{xx}(t, 0) = R_{xx}(t, 0) - \bar{\mu}_x^2(t) = \sigma_x^2(t) \qquad (1\text{-}113b)$$

It is seen that $\sigma_{xy}^2(t)$ is the covariance between $\{{}^k x(t)\}$ and $\{{}^k y(t)\}$ at a given time t and $\sigma_{xx}^2(t)$ is the autovariance of $\{{}^k x(t)\}$ at a given time t.

The random processes $\{^kx(t)\}$ and $\{y^k(t)\}$ are said to be *weakly stationary*, both individually and jointly, if the quantities defined by (1-107) and (1-113a, b) do not depend on t. The random process is said to be *strongly stationary*, both individually and jointly, if all possible statistical quantities of $\{^kx(t)\}$ and $\{^ky(t)\}$ are not affected by a choice of the time t. A random process that is not stationary is called *nonstationary*. Thus under the assumption of stationarity

$$\mu_x = \bar{\mu}_x(t) = \langle ^kx(t)\rangle_e = \text{constant}$$

$$\mu_y = \bar{\mu}_y(t) = \langle ^ky(t)\rangle_\epsilon = \text{constant}$$

$$\rho_{xy}(\tau) = R_{xy}(\tau) - \mu_x\mu_y \tag{1-114}$$

$$\rho_{xx}(\tau) = R_{xx}(\tau) - \mu_x^2$$

where $R_{xy}(\tau)$ is now independent of the choice of t. It is recalled that $\tau = t_2 - t_1$ and can have any finite value (if the sample function is assumed to exist for all time).

Up to now all averages and variances defined in this section were ensemble averages. However, it is also possible to select any two sample functions $^kx(t)$ and $^ky(t)$ from the random processes $\{^kx(t)\}$ and $\{^ky(t)\}$ and average them over time. Such averages are called time or *temporal* averages defined as

$$^k\mu_x = \lim_{T\to\infty} \frac{1}{T}\int_{T/2}^{T/2} {}^kx(t)\,dt \tag{1-115a}$$

$$^k\mu_y = \lim_{T\to\infty} \frac{1}{T}\int_{-T/2}^{T/2} {}^ky(t)\,dt \tag{1-115b}$$

Notice that the means $^k\mu_x$ and $^k\mu_y$ are independent of time since the integrals are definite.

The temporal covariance is defined as

$$^k\rho_{xy}(\tau) = \langle[^kx(t) - {}^k\mu_x][^ky(t+\tau) - {}^k\mu_y]\rangle_T$$

$$= \lim_{T\to\infty} \frac{1}{T}\int_{-T/2}^{T/2}[^kx(t) - {}^k\mu_x][^ky(t+\tau) - {}^k\mu_y]\,dt$$

$$^k\rho_{xy}(\tau) = {}^kR_{xy}(\tau) - {}^k\mu_x{}^k\mu_y \tag{1-116}$$

where

$$^kR_{xy}(\tau) = \langle ^kx(t)^ky(t+\tau)\rangle_T = \lim_{T\to\infty} \frac{1}{T}\int_{-\infty}^{\infty} {}^kx(t)^ky(t+\tau)\,dt$$

Equation 1-116 specializes in an obvious manner to the autocovariance; that is, when $^kx(t) = {}^ky(t)$. A stationary random process is said to be *ergodic* if the ensemble averages (averages over the sample space) are equal to the temporal averages (averages over time of a single record) over a long duration. Hence regardless of which pair of functions are selected from the

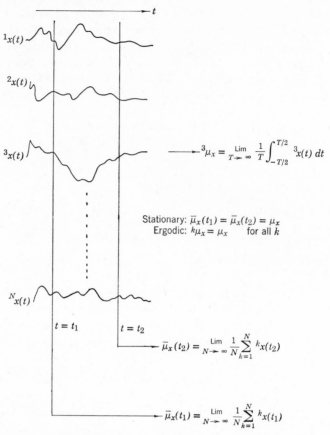

Figure 1-16 Determination of ensemble and temporal means of a random process $\{{}^{k}x(t)\}$.

stationary random process, we have

$$
\begin{aligned}
{}^{k}\mu_x &= \mu_x & {}^{k}\mu_y &= \mu_y \\
{}^{k}\rho_{xy}(\tau) &= \rho_{xy}(\tau) & {}^{k}\rho_{xx}(\tau) &= \rho_{xx}(\tau)
\end{aligned}
\tag{1-117}
$$

where μ_x, μ_y, $\rho_{xx}(\tau)$, $\rho_{xy}(\tau)$ are given by (1-114). These results are summarized pictorially in Figure 1-16. If equation 1-117 holds, then the random processes $\{{}^{k}x(t)\}$ and $\{{}^{k}y(t)\}$ are said to be weakly ergodic. If all the ensemble averages are deducible from the corresponding temporal averages, the random processes are said to be strongly ergodic. It should be realized that an ergodic process must be stationary but the converse is not necessarily true.

Two inequalities concerning the correlation functions are worth noting

$$R_{xx}(0) \geq |R_{xx}(\tau)|$$

$$[R_{xx}(0)R_{yy}(0)]^{1/2} \geq |R_{xy}(\tau)|$$

which are valid for all values of τ. It is easily seen that the latter expression reduces to the former when $^k x(t) = {}^k y(t)$.

It should also be pointed out that since the means of stationary random processes are a constant, independent of time, it would be a trivial assumption if in future chapters we assumed that the means of the random processes were zero. After determining the correlation function one need only to add a constant to obtain the covariance function. Thus in subsequent chapters, when the stationarity assumption is employed the means will often be assumed zero.

Let us now return to our previous discussion of the probability density function. If the random amplitude of any of the member functions of the ensemble at time t_1 is ξ_1, the probability is

$$\mathscr{P}(x_1 < \xi_1 < x_1 + dx_1; t = t_1) = P_{\xi_1}(x_1; t_1)\, dx_1 \qquad (1\text{-}118)$$

If the probability density of the amplitude of an ensemble is the same no matter at what time the observations are made, the ensemble is stationary and (1-118) becomes

$$\mathscr{P}(x_1 < \xi_1 < x_1 + dx_1; t = t_1) = P_{\xi_1}(x_1)\, dx_1 \qquad (1\text{-}119)$$

where $P_{\xi_1}(x_1)$ is independent of t_1. The two-dimensional counterpart to (1-119) is

$$\mathscr{P}_0 = \mathscr{P}(u < \xi < u + du, t = t_1; v < \eta < v + dv, t = t_2)$$
$$= P_{\xi\eta}(u, t_1; v, t_2)\, du\, dv \qquad (1\text{-}120)$$

where $\mathscr{P}_0$ is the probability that at $t = t_1$, ξ is the amplitude found in the interval $(u, u + du)$ and later, at time $t = t_2$, η is found to be in the interval $(v, v + dv)$. If the process is assumed to be stationary, $\mathscr{P}_0$ is independent of the values of t_1 and t_2 and instead just their difference, $\tau = t_2 - t_1$. Then (1-120) becomes

$$\mathscr{P}_0 = P_{\xi\eta}(u, v; \tau)\, du\, dv \qquad (1\text{-}121)$$

Using (1-89b), (1-110), and (1-120) we find that for nonstationary random processes

$$R_{xy}(t_1, t_2) = \int_{-\infty}^{\infty}\int_{-\infty}^{\infty} uv\, P_{\xi\eta}(u, t_1; v, t_2)\, du\, dv \qquad (1\text{-}122)$$

with an obvious specialization to the case when $^k x(t) = {}^k y(t)$. If the process is stationary, (1-122) reduces to

$$R_{xy}(\tau) = \int_{-\infty}^{\infty}\int_{-\infty}^{\infty} uv\, P_{\xi\eta}(u, v; \tau)\, du\, dv \qquad (1\text{-}123)$$

1-5.5   Correlation and Power Spectra

The concept of spectral density has already been introduced in connection with aperiodic functions. However, there are certain mathematical difficulties encountered in extending those results to random function, mainly because the Fourier transform of the sample function does not exist. Consequently, the Fourier integral and the energy spectral density cannot be defined. It is, therefore, useful to introduce the concept of *power* spectral density on the assumption that the average power of the random signal is finite. Hence we consider the following samples of the random processes $\{^k x(t)\}$ and $\{^k y(t)\}$

$$
\begin{aligned}
^k x_T(t) &= {}^k x(t) & t &\leq |T| \\
&= 0 & t &> |T|
\end{aligned}
\tag{1-124a}
$$

and

$$
\begin{aligned}
^k y_T(t) &= {}^k y(t) & t &\leq |T| \\
&= 0 & t &> |T|
\end{aligned}
\tag{1-124b}
$$

Then the Fourier transform of (1-124a, b) exists since they are over a finite time. Thus using (1-48) we obtain

$$
^k X_T(\omega) = \int_{-T}^{T} {}^k x_T(t) e^{-j\omega t} \, dt
\tag{1-125a}
$$

$$
^k Y_T(\omega) = \int_{-T}^{T} {}^k y_T(t) e^{-j\omega t} \, dt
\tag{1-125b}
$$

The average power in the signal is obtained by taking (1-8) to the limit; that is

$$
^k P_{xx_{\mathrm{av}}} = \lim_{T \to \infty} \frac{1}{2T} \int_{-T}^{T} {}^k x_T^2(t) \, dt
\tag{1-126}
$$

and a similar expression for $^k P_{yy_{\mathrm{av}}}$. Using (1-55) we see that (1-126) can be written as

$$
^k P_{xx_{\mathrm{av}}} = \lim_{T \to \infty} \int_0^\infty \frac{|{}^k X_T(\omega)|^2}{T} \, d\omega
\tag{1-127}
$$

Unfortunately (1-127) is an inconsistent estimate of the average power.[*] Hence another approach, and one more suitable from our point of view since it is the way in which it is done experimentally, is to define $^k P_{xx_{\mathrm{av}}}$ in a small band of frequencies, $\Delta\omega$, centered around ω_c. Then, we let

$$
^k P_{xx}(\omega_c, \Delta\omega, T) = \int_{\omega_c - \Delta\omega/2}^{\omega_c + \Delta\omega/2} {}^k \mathscr{G}_{xx}(\omega, T) \, d\omega
\tag{1-128a}
$$

[*] See ref. 9, p. 66 for a discussion of this point.

where

$$^k\mathscr{G}_{xx}(\omega, T) = \frac{|^kX_T(\omega)|^2}{2\pi T} \tag{1-128b}$$

The quantity $^kP_{xx}(\omega_c, \Delta\omega, T)$ is the average power in a record of length $2T$, in a range of frequencies $\Delta\omega$ centered at ω_c. Then the average power over the ensemble is

$$P_{xx}(\omega_c, \Delta\omega, T) = \langle ^kP_{xx}(\omega_c, \Delta\omega, T)\rangle_e = \int_{\omega_c - \Delta\omega/2}^{\omega_c + \Delta\omega/2} \hat{G}_{xx}(\omega, T)\, d\omega \tag{1-129a}$$

wherein

$$\hat{G}_{xx}(\omega, T) = \langle ^k\mathscr{G}_{xx}(\omega, T)\rangle_e \tag{1-129b}$$

The power spectral density $S_{xx}(\omega)$ is determined by

$$G_{xx}(\omega_c) = \lim_{\substack{T \to \infty \\ \Delta\omega \to 0}} \frac{P_{xx}(\omega_c, \Delta\omega, T)}{\Delta\omega} \tag{1-130}$$

Equation 1-130 is the time-averaged power spectral density. It is important to note that $G(\omega) = 2S(\omega)$, $\omega \gtreqless 0$, since (1-130) is a function of only the positive (physically realizable) frequencies. Recall the discussion preceding (1-9).

Equation 1-130 can be extended to define the cross-power spectral density. Let

$$^k\mathscr{G}_{xy}(\omega, T) = \frac{^kX_T^*(\omega)^kY_T(\omega)}{2\pi T} \tag{1-131}$$

then (1-130) becomes

$$G_{xy}(\omega_c) = \lim_{\substack{T \to \infty \\ \Delta\omega \to 0}} \frac{P_{xy}(\omega_c, \Delta\omega, T)}{\Delta\omega} \tag{1-132}$$

where

$$P_{xy}(\omega_c, \Delta\omega, T) = \int_{\omega_c - \Delta\omega/2}^{\omega_c + \Delta\omega/2} \hat{G}_{xy}(\omega, T)\, d\omega \tag{1-133a}$$

and

$$\hat{G}_{xy}(\omega, T) = \langle ^k\mathscr{G}_{xy}(\omega, T)\rangle_e \tag{1-133b}$$

Up to now we have defined correlation and power spectra separately. As in the case with periodic and aperiodic function these two quantities can be related. Limiting ourselves to the stationary random process we have

$$G_{xy}(\omega) = 4\int_0^\infty R_{xy}(\tau) \cos \omega\tau\, d\tau \tag{1-134a}$$

and

$$R_{xy}(\tau) = \frac{1}{2\pi} \int_0^\infty G_{xy}(\omega) \cos \omega\tau\, d\omega \tag{1-134b}$$

since $G_{xy}(\omega)$ and $R_{xy}(\tau)$ are even functions. Equation 1-134a, b constitutes the Wiener-Khintchin theorem for stationary random processes. These results specialize in an obvious manner when ${}^kx(t) = {}^ky(t)$. From the discussion following (1-127) we see that when $\tau = 0$ in (1-134b) $R_{xx}(0)$ is the average power in the random process. For the remainder of the book we limit ourselves to stationary random processes and these latter relations are used frequently.

To illustrate the results of this section, consider the case of four stationary random processes $\{{}^kx(t)\}$, $\{{}^ky(t)\}$, $\{{}^kn(t)\}$, and $\{{}^km(t)\}$, combined to yield two new processes $\{{}^ku(t)\}$ and $\{{}^kv(t)\}$ such that

$$
\begin{aligned}
{}^ku(t) &= {}^kx(t) + {}^kn(t) \\
{}^kv(t) &= {}^ky(t) + {}^km(t)
\end{aligned}
\tag{1-135}
$$

We now determine the cross-correlation between them. Thus from (1-110)

$$
\begin{aligned}
R_{uv}(\tau) &= \langle {}^ku(t){}^kv(t + \tau) \rangle_e \\
&= R_{xy}(\tau) + R_{ny}(\tau) + R_{xm}(\tau) + R_{nm}(\tau)
\end{aligned}
\tag{1-136}
$$

If $[\{{}^kx(t)\}, \{{}^km(t)\}]$, and $[\{{}^ky(t)\}, \{{}^kn(t)\}]$ are statistically independent, then (1-136) reduces to

$$
R_{uv}(\tau) = R_{xy}(\tau) + R_{nm}(\tau)
\tag{1-137}
$$

If, further, ${}^kx(t) = {}^ky(t)$ and ${}^kn(t) = {}^km(t)$, (1-137) becomes

$$
R_{uu}(\tau) = R_{xx}(\tau) + R_{nn}(\tau)
\tag{1-138}
$$

which is the autocorrelation function for the sum of two statistically independent processes. Under these conditions we see that the total autocorrelation function is the sum of the autocorrelation function for each random process. Using the Wiener-Khintchin theorem we see that

$$
G_{uu}(\omega) = G_{xx}(\omega) + G_{nn}(\omega)
\tag{1-139}
$$

Thus the total power spectrum is the sum of the individual power spectrums of the two processes.

Let us return to (1-137) and assume that $\{{}^kn(t)\}$ and $\{{}^km(t)\}$ are statistically independent. Then (1-137) reduces to

$$
R_{uv}(\tau) = R_{xy}(\tau)
\tag{1-140}
$$

Upon comparing (1-140) with (1-138) one can conclude that if the properties of the signal $x(t)$ are known a priori cross-correlation is superior to autocorrelation. This property will be illustrated in Chapter 2. We see that (1-140) behaves like the autocorrelation function except shifted in time an amount τ_0, the delay between the two signals. The Fourier transform of

(1-140) yields the cross power spectrum. Hence

$$G_{uv}(\omega) = G_{xy}(\omega) \qquad (1\text{-}141)$$

Let us now consider several specific examples to illustrate these results. In particular we consider (1-138). We assume that $x(t)$ is a sine wave of frequency ω_0. Its autocorrelation function has already been obtained in (1-40). Thus (1-138) becomes

$$R_{uu}(\tau) = \frac{A_0^2}{2} \cos \omega_0 \tau + R_{nn}(\tau) \qquad (1\text{-}142)$$

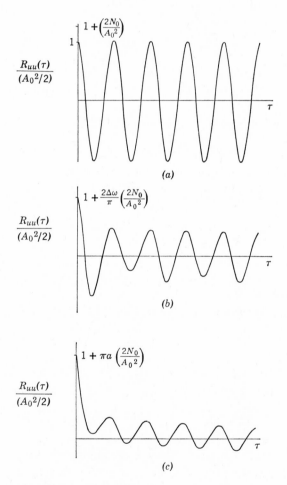

Figure 1-17 Autocorrelation of a sine wave with various amounts of additive noise.

Now consider the case where $\{^{k}n(t)\}$ is noise having the following three power spectral densities.

1. $G_{nn}(\omega) = N_0$. This is the case of broad-band white noise. Its auto-correlation function has been obtained in (1-71). Thus (1-142) becomes

$$R_{uu}(\tau) = \frac{A_0^2}{2} \cos \omega_0 \tau + N_0 \, \delta(\tau) \qquad (1\text{-}143)$$

2. $G_{nn}(\omega) = N_0$, $\omega_0 - \Delta\omega/2 < |\omega| < \omega_0 + \Delta\omega/2$. This is the case of narrow-band white noise. Its autocorrelation function has been obtained in (1-58). Thus (1-142) becomes

$$R_{uu}(\tau) = \left[\frac{A_0^2}{2} + \frac{2N_0}{\pi} \Delta\omega \left(\frac{\sin \Delta\omega\tau}{\Delta\omega\tau} \right) \right] \cos \omega_0 \tau \qquad (1\text{-}144)$$

3. $G_{nn}(\omega) = N_0[1 + a^2\omega^2]^{-1}$. This is the case of shaped white noise. Its autocorrelation function has been obtained in (1-73). Hence (1-142) becomes

$$R_{uu}(\tau) = \frac{A_0^2}{2} \cos \omega_0 \tau + a\pi N_0 e^{-|\tau|/a} \qquad a > 0 \qquad (1\text{-}145)$$

The results given by (1-143) through (1-145) are plotted in Figure 1-17. It is apparent from the figure and the above equations that the broader the bandwidth of the noise is the faster its effects on the autocorrelation function of the desired signal die out. Hence one can conclude that autocorrelation should be performed in the presence of the widest possible noise bandwidth.

Application 1-2   Structure of Turbulence

The flow of fluids or air past a body often creates turbulent flow conditions in the form of turbulent boundary layer near the surface of the body or in the form of wakes in the afterflow of the body. This turbulence is statistical in nature and hence is usually described by a spatial and temporal velocity correlation function. This velocity correlation function gives a measure of the correlation length of the turbulence that, in turn, is a measure of the coherence of the turbulence. This correlation length also yields information about the eddy structure in the turbulent flow.

A block diagram of a typical experiment to cross-correlate the velocity is shown in Figure 1-18a. The velocity probes in this case are hot-wire anemometers. These probes consist of a fine platinum or platinum alloy wire that has a large temperature coefficient of resistivity. The wire is heated by an electric current and partially cooled by the flow of the medium past it. The heat loss causes a decrease of the wire temperature and hence resistance. Usually, the resistance of the wire (temperature) is kept constant by varying the current or voltage. The change in the current or voltage required to

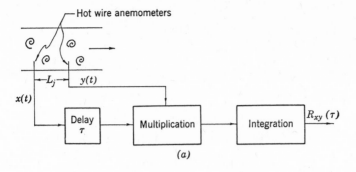

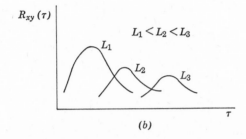

Figure 1-18 (*a*) Functional diagram to determine the correlation properties of turbulence. (*b*) Typical correlograms.

maintain this constant temperature over and above ambient conditions can be calibrated to be proportional to the velocity of the flow past it. Typical cross-correlograms are given in Figure 1-18*b*.

Application 1-3   Recovery of Deterministic Signal in Noise

1. Autocorrelation has proved useful in determining the basic frequency of the brain's alpha rhythm, the dominant rhythm from an adult cortex. With the conventional electroencephalogram (EEG) it is sometimes impossible to measure this basic frequency. A normal alpha rhythm has a smooth correlogram, whereas one with the presence of spikes and other extraneous activity indicate a condition for further medical analysis.

Another use of the EEG is to cross-correlate the output of two sensors, one on the left and one on the right hemisphere of a person's skull. If the maximum delay of the correlogram occurs at approximately zero, it indicates that the electrical activity of both sides of the brain is synchronous. A maximum correlogram at a time different from zero indicates that one side of the brain is not producing comparable rhythmic electrical activity; this sometimes means a tumor is present.

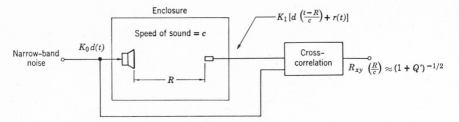

Figure 1-19 Determination of the acoustic ratio of an enclosure.

2. To determine the true randomness of noise one needs only to auto-correlate it. If any peaks other than in the immediate neighborhood of $\tau = 0$ are present, it means that deterministic components to the noise exist and the noise is not truly random.

Application 1-4 Measure of the Acoustic Ratio of an Enclosure

The acoustic ratio, Q', is the ratio of reverberant sound-energy density in an enclosure to the direct sound-energy density. From the knowledge of Q' one can determine the reverberation time of an enclosure. Another method to determine the reverberation time is given in Application 6-4. The values of Q' may range from 0.1 to 1000. It has been shown[*] that Q' can be determined with cross-correlation.

Consider the block diagram shown in Figure 1-19. Narrow-band noise is fed into a loudspeaker that is positioned at one end of the enclosure. A microphone is positioned a distance R away from the loudspeaker. The speed of sound of the air in the enclosure is c. By cross-correlating the input signal to the loudspeaker with that received from the microphone a cross-correlo-gram is obtained. The peak value of the correlogram will occur at $t = R/c$. The value of $R_{xy}(\tau = R/c) = \rho$ yields the acoustic ratio from the relation $Q' = (1/\rho^2) - 1$. This formula is valid only if the cross-correlation between the direct energy and the reverberant energy, $r(t)$, is negligible. This can usually be ensured by using a noise source of sufficient bandwidth.

1-5.6 Errors Associated with Determining Probability Density, Correlation, and Power Spectra[†]

In actual measuring devices the definition given in the Section 1-5.0 are approximate, for in practice one cannot let the bandwidth or amplitude window approach zero as closely as we would like nor can one integrate for an infinite time. Hence we need relations that tell us how much error we are

[*] See ref. 19.
[†] See ref. 2 for the derivation of the formulas presented in this section.

incurring by using finite bandwidths and finite averaging times. The quantity we use was introduced in Section 1-5.3, and is the normalized standard error ϵ given by (1-105). Hence

$$\epsilon = \left(\frac{\sigma}{\mu}\right)_{\text{phenomenon}} = \sqrt{\frac{2}{n_e}} \tag{1-146}$$

We now give the specific form of ϵ for the various functions.

1. Amplitude Probability Density

$$\epsilon \approx A_1[BT\,\Delta x\hat{P}_\xi(x)]^{-\frac{1}{2}} \tag{1-147}$$

where B is the bandwidth, in hertz, of the signal, Δx is the amplitude window (recall Figure 1-11), $\hat{P}_\xi(x)$ is the estimate of the probability density function, A_1 is a constant ($0.15 < A_1 < 0.7$), and T is the averaging time of the device or the length of the record. One sees from (1-147) that the most stringent requirements on the product $BT\,\Delta x$ are demanded by how small a probability density we wish to resolve, or, the equivalent, how many standard deviations from the mean do we still wish to obtain statistically meaningful results. Equation 1-147 is plotted in Figure 1-20 for various choices of $\hat{P}_\xi(x)$.

2. Autocorrelation

$$\epsilon \approx \frac{1}{\sqrt{2BT}}(1 + \alpha^{-2})^{\frac{1}{2}} \tag{1-148}$$

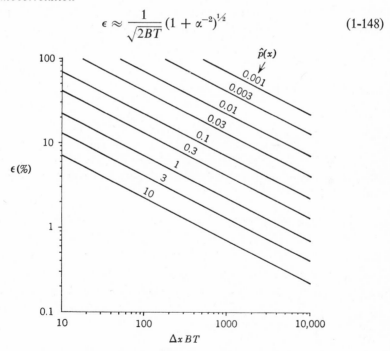

Figure 1-20 Normalized error of the probability density function.

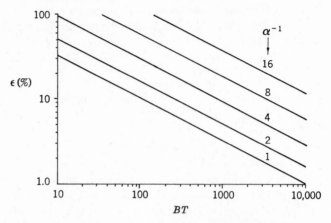

Figure 1-21 Normalized error of the autocorrelation function.

where α is the ratio of smallest amplitude of $R_{xx}(\tau)$ to be resolved to the maximum value $R_{xx}(0)$. It is of interest to note that (1-148) is in agreement with the conclusions drawn from (1-143); namely, that for all else held constant the broader the bandwidth B, the better the resolving power of $R_{xx}(\tau)$ for a given $\tau_{\max}(0 \leq \tau \leq \tau_{\max})$. Equation 1-148 is plotted in Figure 1-21 for various values of α^{-1}.

3. Cross-Correlation

$$\epsilon \approx \frac{1}{\sqrt{2BT}}(1 + \alpha_1^{-1}\alpha_2^{-1})^{\frac{1}{2}} \tag{1-149}$$

where $\alpha_j(j = 1, 2)$ is the smallest amplitude of $R_{12}(\tau)$ to be resolved to the maximum value of $R_{jj}(0)$. The results of Figure 1-21 can be used by simply replacing α^2 by $\alpha_1\alpha_2$.

4. Power Spectral Density

$$\epsilon \approx \frac{1}{\sqrt{B_e T}} \tag{1-150}$$

where B_e is the equivalent bandwidth of the filter. See Section 2-3.1, equation 2-41, for the definition of B_e. Equation 1-150 is plotted in Figure 1-22. Using (1-146) and (1-150) we see that

$$n_e = 2B_e T \tag{1-151}$$

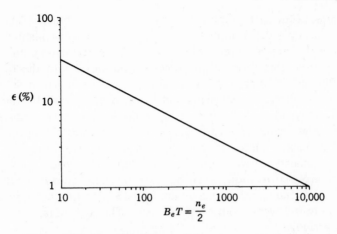

Figure 1-22 Normalized error of the power spectral density.

where, we recall, n_e is the equivalent number of statistical degrees of freedom. Using (1-151) and Figure 1-15, perhaps one can get a better sense of the significance of ϵ.

1-6   Summary

1-6.1   Classification of Signals

Signals are either deterministic or nondeterministic. Deterministic signals can be described by an explicit mathematical relationship as a function of time; nondeterministic signals cannot. Deterministic signals are periodic or aperiodic. Periodic signals repeat themselves in their entirety over some time interval. Aperiodic signals are every other type of deterministic signal. Nondeterministic signals are either stationary or nonstationary. A stationary signal is one in which at least the mean and covariance of the ensemble do not vary with the choice of the time at which the quantities were obtained. Ergodic signals, which are a subclass of stationary signals, are ones in which at least the time average and covariance of a member of the ensemble are equal to the ensemble average and covariance, respectively. Nonstationary signals are all other types of nondeterministic signals.

1-6.2   Periodic Signals

The basic mathematical tool used to describe periodic signals is the Fourier series. The Fourier expansion expresses the shape of the signal in the periodic time interval with a weighted combination of sinusoids whose arguments are harmonics (integer multiples) of the fundamental frequency. The fundamental frequency is the reciprocal of the period. The coefficients multiplying these harmonics are constants, which are independent of time and can be thought of as a transform of the time-domain signal into the frequency domain. The mean-square value of a periodic signal is equal to the sum of the squares of all these coefficients and is the average power in the signal. A plot of the coefficients squared yields a discrete power spectrum for the signal. Correlation between two signals, which involves the operations of time displacement, multiplication, and integration, is a measure of the degree of similarity between the signals. When the signals are different, the term is cross-correlation; when they are the same, autocorrelation. The autocorrelation function is a maximum at zero with this maximum value being equal to the average power of the signal. The correlogram for periodic signals is periodic.

1-6.3   Aperiodic Signals

The basic mathematical tool in the analysis of aperiodic signals is the complex Fourier transform. This transform pair relates the frequency spectrum of the signal to its time-varying wave form and vice versa. The frequency spectrum of an aperiodic signal is continuous whereas for a periodic signal it is a discrete spectra. The Fourier transform of an aperiodic signal multiplied by its complex conjugate yields a real quantity called the spectral energy density of the signal. The integral of the spectral density over all frequencies gives the total energy in the signal. An autocorrelation and cross-correlation function, similar in form to that for periodic signals, exist for aperiodic signals. The correlation function and the spectral density for aperiodic signals form a Fourier transform pair. The autocorrelation function at zero equals the total energy in the signal.

1-6.4   Random Signals

The fundamental quantity used to describe random signals is its probability density function. The probability function for time-varying wave forms shows the relative frequency of occurrence (or the fraction of time spent) of the various instantaneous values of the wave form in a given amplitude interval. Combining a weight function with the probability density and integrating the product over all possible amplitudes gives the expected value of the weight function. In the special case where the weight function is x we obtain the mean value; when it is x^2 we obtain the mean-square value. A measure of the spread of the density function is the variance which is equal to the mean-square value minus the square of the mean. When the mean is zero, the variance is equal to the mean-square value. The square root of the variance is called the standard deviation, which is a measure of the amount of dispersion about the mean. For stationary processes the Weiner-Khintchin theorem states that the correlation function and the spectral density form a Fourier transform pair. The power spectral density for random processes is the average power over a narrow band of frequencies. When two statistically independent random processes are summed, the autocorrelation function for the total process is equal to the sum of the autocorrelation functions of the individual processes. The total power spectral density is the sum of the individual power spectral densities.

References

1. J. S. Bendat, *Principles and Applications of Random Noise Theory*, John Wiley & Sons, Inc., New York, 1958.
2. J. S. Bendat and A. G. Piersol, *Measurement and Analysis of Random Data*, John Wiley & Sons, Inc., New York, 1966.
3. J. S. Bendat, L. D. Enochson, and A. G. Piersol, "Analytical Study of Vibration Data Reduction Methods," NASA No. N64-15529 September 1963.
4. G. R. Cooper and C. D. McGillem, *Method of Signal and System Analysis*, Holt, Rinehart and Winston, New York, 1967.
5. H. Cramer, *Mathematical Methods in Statistics*, Princeton University Press, Princeton, N.J., 1946.
6. S. H. Crandall, Ed., *Random Vibration*, Vol. II, The M.I.T. Press, Cambridge, Mass., 1963.
7. W. B. Davenport and W. L. Root, *Random Signals and Noise*, McGraw-Hill Book Company, Inc., New York, 1958.
8. H. L. Fox, "Probability Density Analyzer," Bolt Beranek and Newman, Inc., Report No. 895 (May 1962).
9. C. W. Horton, Sr., *Signal Processing of Underwater Acoustic Waves*, U.S. Government Printing Office, Washington, D.C., 1969.
10. M. Javid and B. Brenner, *Analysis, Transmission, and Filtering of Signals*, McGraw-Hill Book Company, Inc., New York, 1963.
11. F. H. Lange, *Correlation Techniques*, D. Van Nostrand Co., Inc. Princeton, N.J., 1967.
12. B. P. Lathi, *Signals, Systems and Communication*, John Wiley & Sons, Inc., New York, 1965.
13. Y. W. Lee, *Statistical Theory of Communication*, John Wiley & Sons, Inc., New York, 1960.
14. Y. K. Lin, *Probabilistic Theory of Structural Dynamics*, McGraw-Hill Book Company, Inc., New York, 1967.
15. J. L. Marshall, *Introduction to Signal Theory*, International Textbook Company, Scranton, Pa., 1965.

16. P. F. Panter, *Modulation, Noise and Spectral Analysis*, McGraw-Hill Book Company, Inc., New York, 1965.

17. A. Papoulis, *The Fourier Integral and Its Application*, McGraw-Hill Book Company, Inc., New York, 1962.

18. A. Papoulis, *Probability, Random Variables and Stochastic Processes*, McGraw-Hill Book Company, Inc., New York, 1965.

19. W. R. Stroh, "Direct Measurement of the Acoustic Ratio," *J. Acoust. Soc. Amer.*, **31**, No. 2 (February 1959) pp. 234–238.

20. R. E. Uhrig, *Random Noise Techniques in Nuclear Reactor Design*, Ronald Press Company, New York, 1970.

21. A. A. Winder and C. J. Loda, "Introduction to Acoustical Space-Time Information Processing," ONR Rept. ACR-63, N63-13659 January 1963.

2

FILTERS

2-1 Introduction

Filters are commonplace in electronic instrumentation. They are used in essentially two modes—bandpass and band rejection. In the case of bandpass, the filter permits only those signals whose time-varying wave forms contain frequencies in the band to pass. However, the word "pass" does not necessarily mean to pass unaltered. A filter may intentionally or unintentionally modify the amplitude and/or phase of a signal. In many applications its sole purpose is to "filter out" unwanted signals from the wanted signal. If the unwanted signal is noise, the filter is being used to better the "signal-to-noise"; that is, the extraneous signals interfering with the wanted signal are being (hopefully) minimized while the signal of interest remains relatively unaltered. If the signals are being filtered to determine the frequency components or harmonics of some complicated time-varying wave form, a frequency or harmonic analysis is being performed. If the signal (in the presence of noise) upon passing through the filter, is designed to reach a maximum value in a certain time after having "arrived" at the filter and the shape of the signal need not be preserved, the filter is being used to maximize signal detection in the presence of noise. These are some of the general applications one could perform with a bandpass filter.

A band-rejection filter is exactly the opposite of a bandpass filter. It passes all frequencies outside of the rejection band and greatly attenuates the signals inside of its band. A filter of this type is used when only a very small band of frequencies in the midst of a larger band of frequencies need to be eliminated.

In this chapter the mathematical formulation and analysis of idealized filters in general are discussed first. Then these results are extended to practical filters of various types. In addition, the relationship between bandwidth, averaging time, scan rates, and associated errors are discussed in detail. It should be mentioned, however, that although the words "linear filter" will be used throughout this chapter the results are equally valid if the words "linear device" or "linear system" are substituted.

2-2   Ideal Filters

2-2.1   Definition of Impulse Response and Transfer Function

In general a filter is an electrical network having a pair of input terminals and a pair of output terminals. It may contain resistors, inductors, capacitors, vacuum tubes, relays, or other components. When a time-varying voltage, or "input signal," $s_i(t)$, is applied across the input terminals, a voltage, $s_o(t)$, known as the "output signal," is measured across the output terminals. The filter can be defined by the set of all possible pairs of input and output signals; symbolically this is written as $[s_i(t) \rightarrow s_o(t)]$.

A restriction on all physical filters is that they cannot respond to an input before it begins; that is, if $s_i(t) = 0$ for $t \geqq t_0$, then $s_o(t) = 0$ for $t \leqq t_0$. Another restriction is that the output must remain finite at all times. A filter containing no energy sources is said to be *passive;* otherwise it is called *active.* Only passive filters are considered in this chapter. A filter whose components do not change with time has the property that if $[s_i(t) \rightarrow s_o(t)]$ is an input-output pair, so is $[s_i(t + a) \rightarrow s_o(t + a)]$ for all real values of a. Such a filter is called a time-invariant or stationary filter.

An important class of filters is linear filters. The output of a linear filter is always a linear transformation of the input, which means that if $[s_i^{(1)}(t) \rightarrow s_o^{(1)}(t)]$ and $[s_i^{(2)}(t) \rightarrow s_o^{(2)}(t)]$ are any two pairs of inputs and outputs, then

$$[As_i^{(1)}(t) + Bs_i^{(2)}(t) \rightarrow As_o^{(1)}(t) + Bs_o^{(2)}(t)]$$

is also a possible input-output pair for all values of A and B. The same holds for linear combinations of any number of inputs and outputs.

A linear, stationary filter can be regarded as transforming the input $s_i(t)$ in the following way:

$$s_o(t) = \int_0^\infty K(\tau)s_i(t - \tau)\, d\tau \tag{2-1}$$

The function $K(\tau)$ is known as the *impulse response* of the filter. If the input is a very sharp impulse that can be represented by a delta function at time $t = 0$: $s_i(t) = \delta(t)$ [recall (1-59)], the output of the filter is $s_o(t) = K(t)$, which obviously results from (2-1). Most passive filters contain elements which dissipate energy; hence the response to an impulse eventually dies away so that $K(\tau) \rightarrow 0$ as $\tau \rightarrow \infty$. Since $K(\tau) = 0$ for $\tau < 0$ it is sometimes convenient to let the lower limit of the integral in (2-1) be $-\infty$. In this chapter they are used interchangeably.

The Fourier transform of the impulse response is known as the *transfer function* of the filter designated by $H(\omega)$. Thus

$$H(\omega) = \int_{-\infty}^{\infty} K(\tau)e^{-j\omega\tau}\, d\tau \tag{2-2a}$$

$$K(\tau) = \frac{1}{2\pi} \int_{-\infty}^{\infty} H(\omega)e^{j\omega\tau}\, d\omega \tag{2-2b}$$

In general $H(\omega)$ is a complex quantity. Thus $H(\omega)$ can be expressed in the form

$$H(\omega) = A(\omega)e^{-j\theta(\omega)} \tag{2-3}$$

where $A(\omega) = |H(\omega)|$ and $\theta(\omega) = \tan^{-1}(\operatorname{Im} H(\omega)/\operatorname{Re} H(\omega))$ where Re and Im stand for the real and imaginary parts of $H(\omega)$, respectively. If the input is the complex function of time $e^{j\omega t}$, where ω is the circular frequency, the output of the filter, using (2-1), is

$$s_o(t) = \int_{-\infty}^{\infty} K(\tau)e^{j\omega(t-\tau)}\, d\tau = H(\omega)e^{j\omega t} \tag{2-4}$$

where (2-2a) has been used.

Thus it is seen that a harmonic signal of frequency ω, when passed through a filter of transfer function $H(\omega)$, is modified in amplitude and phase. In addition (2-4) shows that the filter response is completely specified by its sinusoidal steady-state behavior. Using (2-3)

$$s_o(t) = A(\omega)e^{j(\omega t - \theta(\omega))} \tag{2-5}$$

It is noticed that (2-1) is in the form of a convolution integral given by (1-52). Thus if $s_i(\omega)$ and $s_o(\omega)$ are the Fourier transforms of $s_i(t)$ and $s_o(t)$, respectively, then

$$s_o(\omega) = H(\omega)s_i(\omega) \tag{2-6}$$

This result also follows from the definition of a linear filter. Expressing $s_i(\omega)$ in the form

$$s_i(\omega) = B(\omega)e^{j\psi(\omega)} \tag{2-7}$$

where $B(\omega) = |s_i(\omega)|$ and $\psi(\omega) = \tan^{-1}[\operatorname{Im} s_i(\omega)/\operatorname{Re} s_i(\omega)]$. Using (2-3) and (2-7) gives

$$s_o(\omega) = A(\omega)B(\omega)e^{j(\psi(\omega)-\theta(\omega))} \tag{2-8}$$

Equation 2-8 shows that the resulting amplitude spectrum is given by the product of the input spectrum and the filter impulse-response spectrum, while the resulting phase spectrum is given by the difference between the input and impulse-response phase spectra.

2-2.2   Ideal Filters

From the preceding results it is now possible to formulate the requirements of an ideal filter. Namely, if

$$A(\omega) = k = \text{constant} \tag{2-9a}$$

and

$$\theta(\omega) = \omega t_d \tag{2-9b}$$

where t_d is a constant, then (2-6) becomes

$$s_o(\omega) = k \, s_i(\omega) e^{-j\omega t_d}$$

Taking the inverse Fourier transform gives

$$s_o(t) = \frac{k}{2\pi} \int_{-\infty}^{\infty} s_i(\omega) e^{j\omega(t-t_d)} \, d\omega = k \, s_i(t - t_d) \tag{2-10}$$

Thus (2-10) shows that $s_o(t)$ is an exact replica of $s_i(t)$ delayed by t_d seconds. Hence a constant amplitude response and a linear phase response give rise to distortionless transmission of a signal. Also, the slope of the phase response becomes the time delay of the system.

This formulation clearly distinguishes two kinds of distortion in a linear passive network. When k is no longer a constant, the network is said to possess amplitude distortion, and when $\theta(\omega)$ departs from linearity, the network is said to possess phase distortion. In addition to the properties given in (2-9) an ideal filter also has the property of transmitting no frequencies outside of a given band of frequencies.

Using the above properties the following three types of ideal filters exist.

1. Low-Pass

$$
\begin{aligned}
H(\omega) &= ke^{-j\omega t_d} & |\omega| < \omega_c \\
&= 0 & |\omega| > \omega_c
\end{aligned}
\tag{2-11}
$$

2. High-Pass

$$
\begin{aligned}
H(\omega) &= ke^{-j\omega t_d} & |\omega| > \omega_c \\
&= 0 & |\omega| < \omega_c
\end{aligned}
\tag{2-12}
$$

3. Bandpass

$$
\begin{aligned}
H(\omega) &= ke^{-j\omega t_d} & \omega_1 < |\omega| < \omega_2 \\
&= 0 & \omega_1 > |\omega|, |\omega| > \omega_2
\end{aligned}
\tag{2-13}
$$

where ω_c, ω_1, and ω_2 are called the cutoff frequencies. Notice that the cutoff frequencies of an ideal filter define its bandwidth.

2-2.3   Impulse Response of Ideal Filters

The impulse response of a low-pass and bandpass filter will now be found.

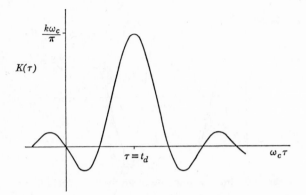

Figure 2-1 Impulse response of an ideal low-pass filter.

1. Low-Pass Filter. Using (2-2b) and (2-11) gives

$$K(\tau) = \frac{1}{2\pi} \int_{-\omega_c}^{\omega_c} k e^{j\omega(\tau - t_d)} \, d\omega = \frac{k\omega_c}{\pi} \frac{\sin \left[\omega_c(\tau - t_d) \right]}{\omega_c(\tau - t_d)} \tag{2-14}$$

Equation 2-14 is shown in Figure 2-1. It is seen that the filter furnishes a response before $t = 0$. This anticipatory response or prediction violates the physics of the problem. Since the analysis is not in error, it must be concluded that the idealized amplitude and phase characteristics given by (2-11) are not possible in a physical network. This is in fact the case. However, a more important conclusion can be drawn; the amplitude and phase characteristics of physical networks are not independent but bear an implicit relation to one another. This relationship, in fact, can be obtained from the Hilbert transform.*

Even though this result is not physically realizable it does give some indication of the type of response one should expect from a realizable filter. If these results are regarded with the understanding that no response is meaningful prior to $t = 0$ and that $\omega_c t_d$ is large, the following conclusion can be drawn: the peak value of the response is proportional to the cutoff frequency ω_c.

2. Bandpass Filter. Using (2-2b) and (2-13) gives

$$K(\tau) = \frac{1}{2\pi} \int_{-\omega_0 - \Delta\omega}^{-\omega_0 + \Delta\omega} k e^{j\omega(\tau - t_d)} \, d\omega + \frac{1}{2\pi} \int_{\omega_0 - \Delta\omega}^{\omega_0 + \Delta\omega} k e^{j\omega(\tau - t_d)} \, d\omega$$

* See, for example, refs. 16 or 17 of Chapter 1.

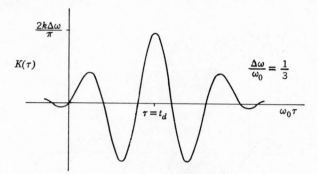

Figure 2-2 Impulse response of an ideal bandpass filter.

where $\Delta\omega = (\omega_2 - \omega_1)/2$ and $\omega_0 = \omega_1 + \Delta\omega$. This integral has already been evaluated in (1-58). Thus

$$K(\tau) = \frac{2k}{\pi} \Delta\omega \cos \left[\omega_0(\tau - t_d)\right] \frac{\sin \left[\Delta\omega(\tau - t_d)\right]}{\Delta\omega(\tau - t_d)} \tag{2-15}$$

This result was shown in Figure 1-8b and is replotted on a different scale in Figure 2-2.

As seen in Figure 2-2, the response is similar to that of the low-pass filter except for the cos $\left[\omega_0(t - t_d)\right]$ term. Again it is seen that the peak response is proportional to the bandwidth of the filter. Equation 2-15 reduces to (2-14) since $\omega_1 = 0$ and $\omega_0 = \Delta\omega = \omega_c/2$. Notice, however, that when $\Delta\omega \to 0$ ($\Delta\omega(\tau - t_d) \ll 1$) we have a very narrow filter, and (2-15) shows that $K(\tau)$ oscillates for a much longer time before dying out. We see then that this delay in decay is related to bandwidth. One should, therefore, expect a narrower band filter to oscillate longer when subjected to a transient loading than a broader band filter.

2-2.4 Correlation and Spectral Density

Let us assume that the input signal to a linear filter is the stationary random process $\{^ks_i(t)\}$ which has an autocorrelation $R_{ii}(\tau)$. The filter has an impulse response $K(\tau)$. To determine the autocorrelation function $R_{oo}(\tau)$ of the output signal we use (1-110) and (2-1) to find

$$R_{oo}(\tau) = \left\langle \int_{-\infty}^{\infty} K(\lambda)^k s_i(t - \lambda) \, d\lambda \int_{-\infty}^{\infty} K(\eta)^k s_i(t + \tau - \eta) \, d\eta \right\rangle_e$$

$$= \int_{-\infty}^{\infty} K(\lambda) \int_{-\infty}^{\infty} K(\eta) R_{ii}(\tau + \lambda - \eta) \, d\eta \, d\lambda \tag{2-16}$$

Equation 2-16 shows the relationship between the autocorrelation function of the input signal to, and the output signal from, a filter whose impulse response is $K(t)$.

The power spectral density of the output signal is obtained from (1-136a). Substituting (2-16) into (1-136a) and performing some manipulations gives

$$G_{oo}(\omega) = |H(\omega)|^2 \, G_{ii}(\omega) \qquad \omega \gtreqless 0 \qquad (2\text{-}17)$$

where $H(\omega)$ is the transfer function of the filter given by (2-2a) and $G_{ii}(\omega)$ is the power spectral density of the input random process which is obtained from the Fourier transform of $R_{ii}(\tau)$. It should be noticed that $|H(\omega)|^2$ is a real, positive quantity, and therefore contains no phase information.

We can now use (1-134b) to obtain another, and perhaps more useful, form for $R_{oo}(\tau)$. Thus substituting (2-17) into (1-134b) we obtain

$$R_{oo}(\tau) = \frac{1}{2\pi} \int_0^\infty |H(\omega)|^2 \, G_{ii}(\omega) \cos \omega\tau \, d\omega \qquad (2\text{-}18)$$

Let us assume that $\langle {}^k x_i(t) \rangle_e = 0$; that is, the mean of the input random process is zero. Then, when $\tau = 0$ in (2-18) we obtain the variance or the average power of the random process at the filter output. Thus

$$\sigma^2 = R_{oo}(0) = \frac{1}{2\pi} \int_0^\infty |H(\omega)|^2 \, G_{ii}(\omega) \, d\omega \qquad (2\text{-}19)$$

This result will be very useful when we talk about effective bandwidths later in this chapter.

Let us now generalize these results to the case of cross-correlation. Consider two stationary random processes $\{{}^k x_i(t)\}$ and $\{{}^k y_i(t)\}$. If we pass $\{{}^k x_i(t)\}$ through a linear filter having an impulse $K_1(t)$ and $\{{}^k y_i(t)\}$ through a linear filter having an impulse response $K_2(t)$, the cross-correlation of the outputs of these two devices will be

$$R_{o_{xy}}(\tau) = \int_0^\infty K_1(\lambda) \int_0^\infty K_2(\eta) R_{i_{xy}}(\lambda + \tau - \eta) \, d\eta \, d\lambda \qquad (2\text{-}20)$$

When $K_1(\tau) = K_2(\tau)$, we have the case wherein two different random processes are passed through identical linear systems. The cross power spectrum, upon substituting (2-20) into (1-134b), yields

$$G_{o_{xy}}(\omega) = H_1(\omega) H_2^*(\omega) G_{i_{xy}}(\omega) \qquad \omega \gtreqless 0 \qquad (2\text{-}21)$$

We now return to (2-1) wherein the input signal is a member of the stationary random process $\{{}^k s_i(t)\}$. Thus (2-1) becomes

$${}^k s_o(t) = \int_0^\infty K(\tau) {}^k s_i(t - \tau) \, d\tau \qquad (2\text{-}22)$$

If we multiply (2-22) by ${}^k s_i(t - \lambda)$ and take the ensemble average of the result, we obtain the very important result

$$R_{io}(\tau) = \int_0^\infty K(t)R_{ii}(\tau - t)\, dt \tag{2-23}$$

where $R_{io}(\tau)$ is the cross-correlation of the input to the linear device with its output. The power spectral density, obtained by substituting (2-23) into (1-134b), yields

$$G_{io}(\omega) = H_{io}(\omega)G_{ii}(\omega) \qquad \omega \gtreqless 0 \tag{2-24}$$

where we have subscripted $H(\omega)$ to indicate that the transfer function is between the points i and o. The significance of this result will be brought out in several applications to follow. Equation 2-24 is a complex quantity, which is often expressed in the form,

$$G_{io}(\omega) = C_{io}(\omega) - jQ_{io}(\omega) \tag{2-25}$$

where $C_{io}(\omega)$ is called the co-spectral density function and $Q_{io}(\omega)$ is the quadrature (quad) spectral density function.

Equation 2-25 can be rewritten as

$$G_{io}(\omega) = A(\omega)e^{-j\varphi(\omega)} \tag{2-26}$$

where $A^2(\omega) = C_{io}{}^2(\omega) + Q_{io}{}^2(\omega)$ and $\varphi(\omega) = \tan^{-1}[Q_{io}(\omega)/C_{io}(\omega)]$. The inverse of (2-26) given by (1-134b) can now be written in the form

$$R_{io}(\tau) = \frac{1}{2\pi} \int_0^\infty A(\omega)e^{j(\omega\tau - \varphi(\omega))}\, d\omega$$
$$= \int_0^\infty A(\omega) \cos \varphi(\omega) \cos \omega\tau\, d\omega + \int_0^\infty A(\omega) \sin \varphi(\omega) \sin \omega\tau\, d\omega \tag{2-27}$$

where we have used the fact that $H_{io}(-\omega) = H_{io}^*(\omega)$ [recall (1-53)ff]. If the output signal ${}^k s_o(t)$ and the input signal ${}^k s_i(t)$ are filtered by an ideal filter of bandwidth $\Delta\omega$ nominally centered at ω before the cross-correlation is performed, (2-27) can be approximated as

$$R_{io}(\tau) \approx A(\omega)[\cos \varphi(\omega) \cos \omega\tau + \sin \varphi(\omega) \cos \omega\tau]\, \Delta\omega \tag{2-28}$$

Using (2-25) and (2-26), (2-28) becomes

$$R_{io}(\tau) \approx [C_{io}(\omega) \cos \omega\tau + Q_{io}(\omega) \sin \omega\tau]\, \Delta\omega \tag{2-29}$$

Recalling (1-132) and the fact that $R_{io}(0) = \langle {}^k s_i(t) {}^k s_o(t) \rangle_e$, (2-29) yields the important result

$$\langle {}^k s_i(t) {}^k s_o(t) \rangle_e = C_{io}(\omega)$$
$$\left\langle {}^k s_i(t) {}^k s_o\left(t + \frac{\pi}{2\omega}\right) \right\rangle_e = Q_{io}(\omega) \qquad \omega - \frac{\Delta\omega}{2} \leqq \omega \leqq \omega + \frac{\Delta\omega}{2} \tag{2-30}$$

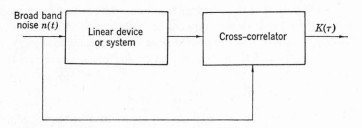

Figure 2-3 Impulse response of a linear system using cross-correlation and broad-band white noise.

where $\tau = \pi/2\omega$ signifies a 90° phase shift. The left-hand side of (2-30) simplifies accordingly if the input signals are ergodic or deterministic. We shall now illustrate these results with a discussion of several techniques.

Application 2-1   Impulse Response of a Linear System

Consider a system shown in Figure 2-3. It is desired that the impulse response of the linear device be obtained. The input signal to the device is broad-band white noise whose autocorrelation function is given by (1-71). This input signal is one input to the cross-correlator. The second input is the output of the linear device. Using (1-71) and (2-23) we see that the output of the correlator is the impulse response of the linear device.

Application 2-2   Measurement of Cross Power Spectrum

Consider the system shown in Figure 2-4. The input to the linear device is passed through a filter of bandwidth $\Delta\omega$ centered at ω. The output of the linear device is also passed through an identical filter. The output from the former filter is connected to two multipliers. The output of the latter filter is connected to one of the multipliers. Another output from this latter filter is phase shifted 90° and connected to the other multiplier. (Both filters are designed to have the same phase characteristics over the frequency range of interest.) The outputs of both multipliers are then averaged (ensemble or

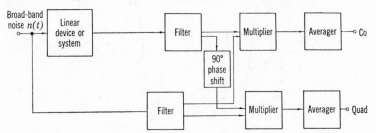

Figure 2-4 Analog determination of the cross-power spectrum.

time). From the output of the averages we obtain the co- and quad spectrum. If the input to the linear device happens to be broad-band white noise then $G_{ii}(\omega) = 2N_0$, which will be a constant over all frequency bands of interest. Thus from (2-24) we determine the amplitude and phase characteristics as a function of ω

$$H_{io}(\omega) = 0.5N_0^{-1}A(\omega)e^{j\varphi(\omega)} \qquad (2\text{-}31)$$

Notice that this technique preserves phase information. Also, it should be noted that this technique is essentially performing the Fourier transform of $K(\tau)$.

The specific applications of this technique are virtually unlimited; however, several examples are minimalization of force transmission between dynamically stressed parts, surface finish analysis, simulation of flow processes around aeroplane surfaces, description of waves in oceanography, and measurement of radiation of electroacoustical transducers.

2-2.5 Basic Properties of Filters

The impulse responses given by (2-14) and (2-15) could now be used in conjunction with (2-1) to solve for the output signal from a low-pass or band-pass filter, respectively, for an arbitrary input signal. It turns out, however, that it is much simpler to work in the frequency domain and use (2-5) and the inverse Fourier transform, (1-49). We now illustrate this latter technique in the following example which has been selected to show the most important effects a filter has on a signal.

Consider a pulsed sinusoid of duration $2t_0$, shown in Figure 2-5, which is described mathematically as

$$\begin{aligned} s_i(t) &= Ae^{j\omega_0 t} & |t| < t_0 \\ &= 0 & |t| > t_0 \end{aligned} \qquad (2\text{-}32)$$

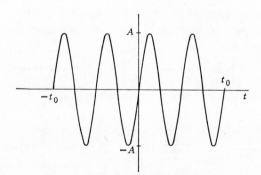

Figure 2-5 Sinusoid of duration $2t_0$.

where ω_0 is the circular frequency of the sinusoid. This particular form was chosen for mathematical convenience. Notice that when $\omega_0 = 0$, (2-32) represents an aperiodic pulse suddenly applied at $t = -t_0$ and suddenly removed at $t = t_0$. Further, we see that when $\omega_0 \neq 0$ and $t_0 \to \infty$, (2-32) represents an harmonic signal. Thus (2-32) has several important special cases which we will examine later. The Fourier transform of (2-32) is [recall (1-48)]

$$s_i(\omega) = A[e^{j(\omega_0 - \omega)t_0} - e^{-j(\omega_0 - \omega)t_0}][j(\omega_0 - \omega)]^{-1} \qquad (2\text{-}33)$$

If we let this signal pass through an ideal bandpass filter given by (2-13), the output signal from the filter is [recall (2-6)]

$$s_o(\omega) = kA\{\exp j[\omega_0 t_0 - \omega(t_0 + t_d)]$$
$$- \exp j[-\omega_0 t_0 + \omega(t_0 - t_d)]\}[j(\omega_0 - \omega)]^{-1} \qquad \omega_1 < |\omega| < \omega_2$$
$$= 0 \qquad \text{otherwise} \qquad (2\text{-}34)$$

where $\omega_1 < \omega_0 < \omega_2$. Substituting (2-34) into (1-49), noting that in the range of negative frequencies the integrals will contribute very little as opposed to those in the neighborhood of ω_0, and letting $(\omega_2 - \omega_1)/2 = \Delta\omega$ and $\omega_0 + \Delta\omega = \omega_2$, we find that

$$s_o(t) = Ak\left\{1 + \frac{1}{\pi} \text{ Si } [\Delta\omega(t' - t_d)] + \frac{1}{\pi} \text{ Si } [\Delta\omega(t' - t_d - 2t_0)]\right\}$$
$$\times \cos (t' - t_0 - t_d) \qquad (2\text{-}35)$$

where Si (y) is an abbreviation for the sine integral function

$$\text{Si } (y) = \int_0^y \frac{\sin x}{x} \, dx \qquad (2\text{-}36)$$

and $t' = t + t_0$.

Equation 2-36 consists of two parts; the carrier and the envelope. The envelope also consists of two parts which are linear sums: the first is the envelope caused by the start of the signal and the second is the envelope caused by the abrupt cessation of the signal. Since the starting and stopping have the same properties only shifted in time an amount $2t_0$, only the starting contribution will be discussed. In addition, it is again noted that if ω_0 is set equal to zero, the resulting equation would be that for a single rectangular pulse of duration $2t_0$. Hence the shape of the pulse is really immaterial, for all shapes would exhibit qualitatively the properties discussed below. It can be shown that the result of (2-36) degenerates to that of a low-pass filter by simply replacing $\Delta\omega$ by ω_c. It should be mentioned at this point, however, that $\Delta\omega/2\pi = B/2$ is half the bandwidth in hertz of the positive frequencies of the bandpass filter. Thus the realizable bandwidth, W, of positive frequencies is $W = B$. For the low-pass filter, however, we have that the realizable bandwidth of positive frequencies is $W = B/2$ (since $\Delta\omega/2\pi = \omega_c/2\pi = B/2$). These results indicate that a low-pass filter, with realizable bandwidth from 0 to W Hz, has twice the effective bandwidth as a bandpass filter whose positive bandwidth is W Hz. Considered in a different way, we can say that noise which is confined to a positive bandwidth B Hz in a bandpass filter will contain the same average power as its translated counterparts which are confined to a bandwidth from 0 to $B/2$ Hz in a low-pass filter, the negative portion from $-B/2$ to 0 apparently "folding" over.

We shall now verify this last statement. If we pass white noise through an ideal low-pass filter of bandwidth $-B/2 < \omega/2\pi < B/2$ and autocorrelate the output, (2-18) gives

$$R_{oo}(\tau) = \frac{N_0}{\pi} \int_0^{B/2} \cos \omega\tau \, d\omega = \frac{N_0 B}{2\pi} \frac{\sin (B\tau/2)}{B\tau/2}$$

whereas the autocorrelated output of an ideal bandpass filter of bandwidth B is

$$R_{oo}(\tau) = \frac{N_0}{\pi} \int_{\omega_0-B/2}^{\omega_0+B/2} \cos \omega\tau \, d\omega = \frac{N_0 B}{\pi} \frac{\sin (B\tau/2)}{B\tau/2} \cos \omega_0\tau$$

The average power is obtained when $\tau = 0$. Hence we see that the average power through the low-pass filter is one half the average power through the bandpass filter, which supports the statement in the paragraph above.

Returning to (2-35) and examining only the starting portion of the output signal, we have

$$s_o(t) = Ak\left\{\frac{1}{2} + \frac{1}{\pi} \text{Si} \left[\Delta\omega(t' - t_d)\right]\right\} \cos \left[\omega_0(t' - t_0 - t_d)\right] \qquad (2\text{-}37)$$

This result is shown in Figure 2-6. The time t_d which has been called the

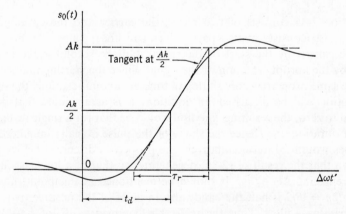

Figure 2-6 Initial response of an ideal bandpass filter to a suddenly impressed voltage.

delay time is seen to be [from (2-37)] that time at which the envelope of the output signal reaches half its final value, in this case $Ak/2$. If at this point a tangent is drawn to the curve (envelope), then the time interval between the intercepts of this tangent with the time axis and the final value of $s_o(t)$ represents fairly well what may be called the build-up time of the filter or the rise time τ_r. Thus

$$\tau_r = \frac{\pi}{\Delta\omega} = \frac{1}{B} \qquad (2\text{-}38)$$

where B is the bandwidth in hertz and we have used the fact that the slope of Si (0) = 1. Thus whereas the time delay is determined by the slope of the phase characteristic alone, the time of build-up is inversely proportional to the filter bandwidth. If the filter is a low-pass filter we have that $\tau_r = 1/2B$.

Another definition is usually employed for τ_r. In electronic circuits it is customary to define τ_r as the time it takes for the response to rise from 10 to 90% of its final value. Using a table for Si (y), it is found that

$$\tau_r = \frac{0.88}{B}$$

for a bandpass filter and $\tau_r = 0.44/B$ for a low-pass filter. In any case, regardless of which definition is used, it can be concluded that

$$\tau_r B = \text{constant} \qquad (2\text{-}39)$$

Equation 2-39 indicates that an infinitely selective system $(B \to 0)$ does not respond to amplitude variations at all. In addition it should be noted from (2-36) and Figure 2-6 that a filter must not only respond to the carrier frequency, $\cos \omega_0 t$, but also to a band of frequencies whose width is determined by the *rate* at which the amplitude of the response is to follow the corresponding amplitude variations in the carrier.

To see a direct consequence of the above result consider the case of a two pulse group wherein the pulse duration is equal to T. When $B \gg 1/T$ the resulting output signal is that shown in Figure 2-7a. When $B \ll 1/T$ the output signal is that shown in Figure 2-7b. In other words, when $\tau_r \approx 1/B \gg T$, the filter (linear system) does not have enough time to respond and, therefore, the amplitude of the output signal is much less than it should be. This is simply a restatement of the result given by (2-15) which showed that the amplitude of the impulse response of the filter is proportional to the bandwidth. In this case the bandwidth should be approximately equal to $1/T$; since it is much less than this value, the filter can not reach its maximum value.

As a final remark it is interesting to note that as $t_0 \to \infty$, the pulse becomes a steady-state signal. Thus

$$\lim_{t_0 \to \infty} s_o(t) = kA \cos [\omega_0(t - t_d)]$$

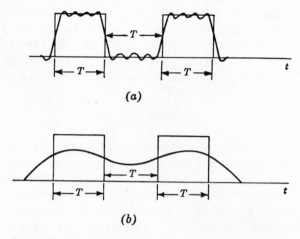

(a)

(b)

Figure 2-7 Output signal of a two-pulse group through a bandpass filter where (a) $B \gg 1/T$ and (b) $B \ll 1/T$.

It is interesting to note that even though the output signal is the same form as the input signal it is still delayed by an amount t_d, which agrees with (2-10).

2-3   Nonideal (Practical) Filters

2-3.1   Introduction

The preceding section was concerned with the ideal, but physically un-realizable filters. Although the results obtained in that section cannot be duplicated in the laboratory one should correctly expect that the properties illustrated by the ideal filter will, at least qualitatively, hold for the nonideal (practical) filter. In Figure 2-8 the exaggerated response of a nonideal bandpass filter and an ideal bandpass filter are shown. It is seen that in the practical filter there is no precise cutoff frequency or center frequency. In addition, the area contained by the ideal filter and that contained by the nonideal one are different, that is, they pass different amounts of power. Hence different means are needed to define a practical filter than those used to define an ideal filter. Recall, that an ideal filter is defined by its cutoff frequencies only, since the amplitude is a constant between these frequencies. In a nonideal filter the amplitude is not a constant but usually varies with frequency. Thus to describe a nonideal filter in a meaningful manner its complete amplitude frequency response, $|H(f)|$, must be known. This information is not always available. Instead, it is usually sufficient to state the amount of deviation, d, from some mean value and the end frequencies in which this d remains valid. It is seen from Figure 2-8 that these limits can

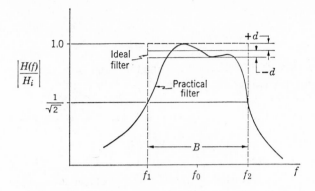

Figure 2-8 Frequency response of an ideal bandpass filter and a practical bandpass filter.

assume radically different values. Thus still another definition is required: the cutoff frequency of a practical filter. The cutoff frequency is defined as those frequencies at which the response $|H(\omega)|$ has decreased to $1/\sqrt{2}$ its mean value, this mean value being the same as that used to define d. This value of $1/\sqrt{2}$ when expressed in terms of decibels (see Appendix I) is called the -3-dB point or half-power point of the filter. Thus in well designed filters of appreciable bandwidth, the value of d should be much less than ±3-dB and should have frequency limits as close as possible to those of the -3-dB points. It will be shown, however, that this requirement can be relaxed for very narrow-band filters.

In addition, the rate at which the filter attenuates the amplitude of frequencies on either side of the cutoff frequencies is also an important factor. This rate is given in terms of attenuation per frequency band. A convenient frequency band is that of an octave. An octave is defined as the doubling (or halving) of a frequency. Hence if f_2 is the upper cutoff frequency, then $2f_2$ is an octave higher. The unit, then, to describe the rate of attenuation is decibels per octave. This slope of the filter attenuation is sometimes called the filter "skirt." It should be pointed out that when one talks about slope over an octave, a straight line between the intercepts of the attenuation curve and the frequencies f_2 and $2f_2$ is implied. However, this is almost never the case, and by stating that the rate of attenuation is X dB/octave usually means that if one goes an octave away from the cutoff frequency, f_2, the attenuation will be $-X$ dB from that value at f_2.

Another descriptive term used to describe a filter is its Q or quality factor, which is the ratio of the center frequency of the filter to its -3-dB bandwidth.

Thus

$$Q = \frac{f_0}{B} \tag{2-40}$$

where B is the -3-dB bandwidth of the filter and f_0 its center frequency.

As mentioned earlier the amount of power transmitted by an ideal and nonideal filter may not be the same. To measure the difference between these two filters a term called the *effective* or *noise bandwidth*, B_e, is defined as

$$B_e = |H_i|^{-2} \int_0^\infty |H(f)|^2 \, df \tag{2-41}$$

where $H(f)$ is the transfer function of the practical filter as a function of frequency (hertz), and $|H_i|$ is the maximum value of $H(f)$ and is a constant for a given filter. See Figure 2-8. The effective or noise bandwidth is then the bandwidth of a hypothetical rectangular (ideal) filter which would pass a signal with the same mean-square value as the actual filter when the input is white noise.

To illustrate (2-41) we consider a filter whose response is given by (1-74), which can be written as

$$\frac{|H(f)|^2}{|H_i|^2} = \left[\frac{\omega^2}{{\omega_0}^2} + Q^2 \left(1 - \frac{\omega^2}{{\omega_0}^2} \right) \right]^{-1} \tag{2-42}$$

and wherein we have used the fact that if $b/a \ll 1$, $2b \simeq \omega_0/Q = 2\pi B$.

Substituting (2-42) into (2-41) and integrating yields

$$B_e = \frac{\pi B}{2}$$

Thus for a simple resonant filter given by (2-42) the effective bandwidth or the bandwidth of the ideal filter would be 57% greater than the -3-dB bandwidth of the practical filter itself. In other words, the practical filter passes 57% more energy than would the ideal filter shown in Figure 2-8.

Another example of the significance of the effective bandwidth has been obtained by Sepmeyer* for the nonideal (but still idealized) bandpass filter shown in Figure 2-9. Referring to the figure, f_1 and f_2 are the cutoff frequencies, A the selectivity or skirt of the filter in decibels per octave and $(\pm)\, m$ the slope of the signal spectrum. Hence we consider here the effect of a nonwhite noise spectrum on a more realistic filter model compared to that of the ideal bandpass filter. From (2-18) we see that the percentage error, H_0, between the ideal filter and the model shown in Figure 2-9 is

$$H_0 = (C - 1) \times 100\% \tag{2-43a}$$

* See ref. 11.

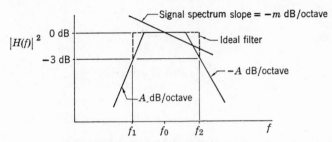

Figure 2-9   Model of a bandpass filter.

where

$$C = \frac{\displaystyle\int_0^\infty |H(f)|^2 \, G_{ii}(f) \, df}{|H_i|^2 \displaystyle\int_{f_1}^{f_2} G_{ii}(f) \, df} \tag{2-43b}$$

and $G_{ii}(f)$ is the spectral density of the input signal denoted graphically as having a slope ($\pm$) m dB/octave in Figure 2-9. Solving for C in (2-43b), it is found that

$$C = \frac{A_0 A}{D - 1}\left[\frac{D}{A_0^2}(A - m - 3)^{-1} - (A + m + 3)^{-1}\right] \tag{2-44a}$$

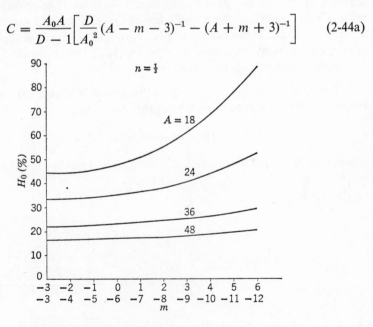

Figure 2-10   Error in the filter of Figure 2-9 as a function of m, A, and n compared to an ideal filter.

where $m \neq -3$ and

$$C = 1 + \frac{2.656}{An} \tag{2-44b}$$

when $m = -3$. The quantities A_0 and D are given by

$$A_0 = 2^{(m+3)/A}$$
$$D = 2^{n(m+3)/3}$$

where n is the number of octaves (see Section 2-4.2). These results are shown in Figure 2-10 for various combinations of n, m, and A. (The value of n should be greater than $\frac{1}{4}$.) It is interesting to note that for the case of white noise ($m = 0$) where $C = B_e$ we find for $n = \frac{1}{3}$ ($\frac{1}{3}$-octave filter) that $H_0 = 35\%$ when $A = 24$ dB/octave and $H_0 = 17\%$ when $A = 48$ dB/octave. This example illustrates a simple fact: the better the selectivity of the filter the closer the practical filter approximates the ideal filter and, therefore, the more accurate the energy in that band can be measured.

2-3.2 RC Filters

At this point it is worthwhile to examine two of the simplest types of practical filters, the low-pass and high-pass RC filter. The R stands for the resistance and the C for the capacitance of the passive elements from which the filter is made. These particular filters are selected because they are used extensively in many electronic devices and will, therefore, be referred to frequently in the subsequent chapters.

1. Low-Pass RC Filter. A low-pass RC circuit is shown schematically in Figure 2-11a. The transfer function of this circuit is

$$H(\omega) = (1 + j\omega\tau_0)^{-1} \tag{2-45}$$

where $\tau_0 = RC$ is the time constant of the circuit. Thus the amplitude density

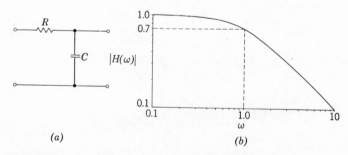

(a)

(b)

Figure 2-11 (a) Low-pass RC filter and (b) its corresponding amplitude spectrum.

spectrum of the filter is

$$S(\omega) = |H(\omega)|^2 = (1 + \omega^2\tau_0^2)^{-1} \tag{2-46}$$

which is plotted in Figure 2-11b.

It is seen from Figure 2-11b that when $\omega = 1/\tau_0$ the spectrum has decreased to one half its maximum value. This corresponds to the -3-dB point. The cutoff frequency, therefore, is given by

$$f_c = \frac{1}{2\pi RC} \tag{2-47}$$

When $\omega \ll 1/\tau_0$, the filter passes signals with almost no amplitude attenuation. The phase angle, given by (2-3), is

$$\varphi(\omega) = \tan^{-1}(\omega\tau_0) \tag{2-48}$$

When $\omega \ll 1/\tau_0$, (2-48) becomes $\varphi(\omega) = \omega\tau_0$ which corresponds to the linear phase relationship of an ideal filter. In addition, the amplitude of the filter response is almost a constant. Thus a low-pass RC filter acts like an ideal filter when $\omega\tau_0 \ll 1$.

If one uses equation 2-4 and 2-45 the following result is obtained:

$$s_0(t) = \frac{e^{j\omega t}}{1 + j\omega\tau_0}$$

Now, when $\omega\tau_0 \gg 1$,

$$s_0(t) \approx \frac{e^{j\omega t}}{j\omega\tau_0} \tag{2-49}$$

But $s_i(t) = e^{j\omega t}$ and, consequently, a low-pass filter acts as an *integrator* for $\omega\tau_0 \gg 1$ since

$$\int s_i(t)\, dt = \int e^{j\omega t}\, dt = \frac{e^{j\omega t}}{j\omega} \sim s_0(t)$$

We notice several interesting properties of integration of a sinusoid by a low-pass filter. First the integration introduces a $90°$ phase shift, and secondly, it decreases the signal amplitude by an amount $\omega\tau_0$. Another important feature of the integration is to notice that the integration time τ_0 will, by the fact that $\omega\tau_0 \gg 1$ for the low-pass filter to act as an integrator, be very much larger than the period of the lowest frequency sinusoid it can integrate. This guarantees that the lowest frequency sinusoid will be integrated over many periods. Although (2-49) was obtained for a single frequency, the integration property of the low-pass filters holds for periodic signals, and aperiodic and random signals of sufficiently long (compared to the "build-up" time of the filter) duration.

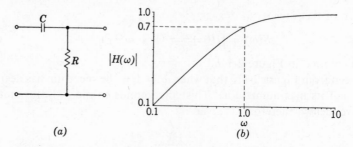

Figure 2-12   (a) High-pass RC filter and (b) its corresponding amplitude spectrum.

Returning to Figure 2-11b we see that the selectivity of a simple RC filter is -6-dB/octave. If we were to add another identical RC filter (without interaction) to the single filter, it is an easy matter to show that the selectivity is -12-dB/octave. Further, one can show that the equivalent bandwidth for the single section is $B_e = 1/4\tau_0$ and that of the two-section filter $B_e = 1/8\tau_0$. Corresponding to each of these filters, the 10 to 90% rise times are $\tau_{r_1} = 2.2\tau_0$ and $\tau_{r_2} = 3.4\tau_0$, respectively.

2. High-Pass RC Filter.   A high-pass RC circuit is shown schematically in Figure 2-12a. The transfer function of this circuit is

$$H(\omega) = \frac{j\omega\tau_0}{1 + j\omega\tau_0} \tag{2-50}$$

Thus

$$S(\omega) = |H(\omega)|^2 = \frac{\omega^2\tau_0^{\,2}}{1 + \omega^2\tau_0^{\,2}} \tag{2-51}$$

Equation 2-51 is plotted in Figure 2-12b.

In this case it is seen that the -3-dB point is again at the point where $\omega = 1/\tau_0$, and (2-47) is also valid here. When $\omega\tau_0 \gg 1$, the filter passes signals with almost no amplitude attenuation. The phase angle is given by

$$\theta(\omega) = \tan^{-1}\left(\frac{1}{\omega\tau_0}\right)$$

When $\omega\tau_0 \gg 1$, $\theta(\omega) \approx 1/\omega\tau_0 \ll 1$, which shows that the phase shift approaches zero as ω increases. If one again uses (2-4) with (2-50), the following is obtained:

$$s_0(t) = \frac{j\omega\tau_0 e^{j\omega t}}{1 + j\omega\tau_0}$$

Now, when $\omega\tau_0 \ll 1$,

$$s_0(t) \approx j\omega\tau_0 e^{j\omega t} \tag{2-52}$$

But $s_i(t) = e^{j\omega t}$ and, consequently, a high-pass filter essentially acts as a *differentiator* when $\omega\tau_0 \ll 1$ since

$$\frac{d}{dt} s_i(t) = \frac{d}{dt} e^{j\omega t} = j\omega e^{j\omega t} \sim s_0(t)$$

It is seen from (2-52) that differentiation also decreases the magnitude of the signal since $\omega\tau_0 \ll 1$ by assumption. Although (2-52) was presented for a sinusoid, it can be extended to all types of signals. As a final remark we note that both mathematically and in practice integration is a smoothing process whereas differentiation accentuates the abrupt changes in the signal. Therefore, if one has a signal in the presence of noise, differentiation accentuates the noise, whereas integration will tend to lessen its effect by smoothing (averaging) the noise.

Many other configurations are possible using these passive elements. Numerous examples are given in Ref. 12, pp. 6-32–6-42.

2-4   Types of Practical Filters

2-4.1   General Classification

Filters can be classified as either a constant percentage bandpass filter, a constant bandwidth filter, or a variable bandwidth filter. These different filters are classified by the relationship of their center frequency to their -3-dB cutoff frequencies. The center frequency is that frequency which is the geometric center between the -3-dB frequencies. Let the center frequency of the filter be denoted by f_0 and the upper and lower cutoff frequencies by f_2 and f_1, respectively. Then a constant percentage bandwidth filter is one in which the bandwidth is some percentage of f_0. A constant bandwidth filter of bandwidth B is one in which $f_2 = f_0 + \alpha B/2$ and $f_1 = f_c - \beta B/2$, where α and β are constants such that f_0 is the geometric mean of the band. A variable bandpass filter is one in which f_2 and f_1 can be selected (within the limits of the filter). In this case the center frequency in general changes with the choice of f_1 and f_2.

These filters are sometimes further classified depending on how they are actually constructed. If a frequency range is divided into N frequency bands (not necessarily equal) and if a filter is constructed so that it consists of N filters each of which is responsive to frequencies only in that band, the filter is termed a *contiguous* filter. On the other hand, if a filter consists of only one filter whose center frequency is continuously variable through the frequency range, the filter is termed a *continuous* filter.

2-4.2 Octave Filters

An octave filter is a constant percentage bandwidth filter in which

$$f_2 = kf_0 \qquad k > 1 \tag{2-53}$$
$$f_1 = k^{-1}f_0$$

such that

$$f_2 = 2^n f_1 \tag{2-54}$$

where n is the number of octaves. The value of n can be an integer or a fraction, if the filter is a one octave filter $n = 1$, or if the filter is a third octave filter $n = \frac{1}{3}$, and so on. From (2-53) and (2-54) it is easy to show that

$$k = 2^{n/2} \tag{2-55}$$

Also from (2-53) it can be shown that $k = \sqrt{f_2/f_1}$, and thus

$$f_0 = \sqrt{f_2 f_1} \tag{2-56}$$

Hence f_0 is called the geometric mean of the frequency band.

The bandwidth of the filter, B, is

$$B = f_2 - f_1 = f_0(k - k^{-1}) = \frac{f_0}{Q} \tag{2-57}$$

wherein (2-40) has been used. Therefore, the n-octave filter is a constant percentage bandwidth filter in which the percentage of f_0, say β, is

$$\beta = k - k^{-1} = (2^n - 1)2^{-n/2} \tag{2-58}$$

For a one octave filter, $n = 1$ and $\beta = 0.707$; for a third octave filter $n = \frac{1}{3}$ and $\beta = 0.231$.

Octave filters are usually contiguous filters designed to operate over a large frequency range (20 to 40,000 Hz). They are extremely useful in area of acoustic measurements, and in subsequent chapters several specific applications involving their use are given.

2-4.3 Frequency Analyzers

Frequency analyzers are either narrow-band constant percentage continuous filters or narrow-band constant bandwidth filters. Consider first the constant percentage analyzer. As stated before, a constant percentage bandwidth filter is one whose bandwidth is defined by some percentage, β, of the center frequency. Thus

$$B = \beta f_0 = \frac{f_0}{Q} \tag{2-59}$$

To determine the cutoff frequencies of this type of filter we use (2-58) and solve for k. Thus

$$k = \frac{\beta}{2} + \left[\left(\frac{\beta}{2}\right)^2 + 1 \right]^{\frac{1}{2}} \tag{2-60}$$

since $k > 1$. To convert this to octaves we use (2-55) to yield

$$n = 6.45 \log_{10} \left[\frac{\beta}{2} + \frac{(\beta^2 + 4)^{\frac{1}{2}}}{2} \right] \tag{2-61}$$

If $\beta = 0.06$ (i.e., a 6% analyzer), (2-61) gives $n = 0.0825$ or less than one tenth of an octave.

The cutoff frequencies in the general case are given by (2-53) with k given by (2-60). For a 6% analyzer $f_2 = 1.03 f_c$ and $f_1 = 0.97 f_c$. Thus it can be seen that for very narrow bandwidths the geometric mean and the arithmetic mean are the same for all practical purposes.

Let us now consider the constant bandwidth analyzer. In this case the bandwidth is given by (2-57). Replacing β by B/f_0 in (2-60) and (2-61) the number of octaves and its upper and lower cutoff frequencies can readily be determined. It is seen that if $B/f_0 \ll 1$, the geometric and arithmetic center frequencies are again the same for all practical purposes.

2-4.4 Tracking Filter

A constant bandwidth filter could be constructed in a contiguous manner as are octave filters. However, this is impractical if the bandwidth of the filter, B, is very small. Instead, a single constant bandwidth filter can be used in conjunction with a process called heterodyning. Heterodyning is an operation whereby a given frequency spectrum is translated from one region of frequency to another. From the definition of heterodyning one takes the signal $f(t) = k_1 \cos \omega_0 t$ and translates it into another signal $f_1(t) = k_2 \cos \omega_1 t$, where $\omega_0 \neq \omega_1$. The heterodyning technique can be used to construct a very narrow constant bandwidth filter operable over a wide frequency range. Such a filter is called a tracking filter. In addition, as we will see later, heterodyning can be used to speed up analysis time.

To understand the process of heterodyning and its usage in a tracking filter consider first the trignometric identity

$$\cos x \cos y = \tfrac{1}{2} [\cos (x - y) + \cos (x + y)] \tag{2-62}$$

We will now use this result by considering the simplified functional diagram of a tracking filter shown in Figure 2-13a. For simplicity we first let the input signal be $f_i(t) = k_0 \cos \omega_0 t$, a sinusoid of frequency ω_0. This input signal is assumed to be the output signal from some device to which a signal of the same frequency, but perhaps different phase, has been applied. We denote this latter signal $f_0(t + \varphi)$ and call it the reference signal. The reference

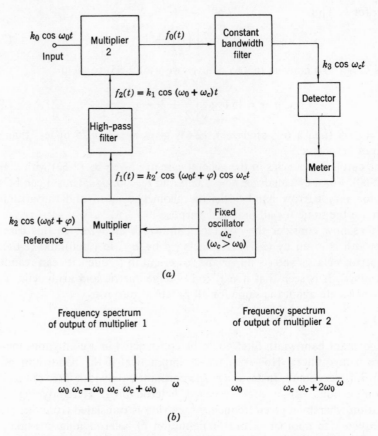

(a)

Frequency spectrum
of output of multiplier 1

Frequency spectrum
of output of multiplier 2

(b)

Figure 2-13 *(a)* Tracking filter and *(b)* the frequency spectrum from the various multipliers.

signal is fed to multiplier 1 where it is multiplied with an internally generated sinusoid of frequency ω_c. Using (2-62) we see that the output of multiplier 1 consists of two sinusoids, one of frequency $\omega_0 + \omega_c$ and the other of frequency $\omega_c - \omega_0 (\omega_c > \omega_0)$. See Figure 2-13b. If we now pass this signal through a high-pass filter which greatly attenuates all frequencies $\leqq \omega_c$, the output signal from the filter will be $f_1(t) = k_1 \cos [(\omega_0 + \omega_c)t + \varphi_1]$ where φ_1 is the phase lag through the filter plus the original phase lag φ. If we multiply $f_1(t)$ with $f_i(t)$, the output signal, using (2-62), becomes

$$f_2(t) = k_0 k_1 \{\cos (\omega_c t + \varphi_2) + \cos [(\omega_c + 2\omega_0)t + \varphi_3]\} \qquad (2\text{-}63)$$

We see that the first term on the right-hand side of (2-63) is always a sinusoid of frequency ω_c whose amplitude is proportional to the amplitude of $f_1(t)$, the input signal. The second term in (2-63) is always at a frequency greater than ω_c. Hence if we place a filter of bandwidth $2\pi B$ centered at ω_c, we can determine the amplitude of $f_1(t)$ provided that $\omega_0 > 2\pi B$ and k_1 is known. The former requirement specifies the lower usable frequency limit and the latter is a calibration constant that is given by the manufacturer such that the meter reads directly the amplitude k_0. The function of the detector is given in Chapter 4.

To extend the above to random signals we note that of all the individual frequency components in the random signal only those components that are in the band $\omega_0 - \pi B < \omega < \omega_0 + \pi B$ will pass through the narrow-band filter. Depending on the detector used the meter records the appropriate amplitude. In certain applications the tracking filter can be thought of as a tuned voltmeter.

The heterodyning technique can also be used to translate down in frequency so that one requires only a low-pass filter to perform the filtering. Let the input signal be at a frequency ω_0 and a reference oscillator signal at ω_c. From (2-62) we see that at the output of the multiplier a difference frequency $\omega_c - \omega_0$ exists. If ω_c is chosen such that $\omega_c = \omega_0 \pm \Delta\omega/2$, then the difference frequency varies from $-\Delta\omega/2$ to $+\Delta\omega/2$. If we use a low-pass filter of bandwidth $\Delta\omega/2$ we see from the discussion at the end of Section 2-2.5 that the bandpass equivalent of the low-pass filter is $\Delta\omega$. Therefore, we have used a low-pass filter as a bandpass filter. Notice, however, that the cutoff frequency of this type of filter can be made extremely narrow and by its very nature it will be perfectly symmetrical as an equivalent bandpass filter. This type of technique has several interesting uses which are described in Chapter 4.

As a final remark it should be mentioned that a different measure of the selectivity of a constant bandwidth filter must be used if $B/f_c \ll 1$, since it would be meaningless to talk about attenuation in decibels per octave. Thus a narrow-band constant bandwidth filter is usually described by a quantity called its *shape factor* which is the amount of attenuation per doubling of bandwidth or the number of decibels per bandwidth octave. An alternate definition of shape factor is sometimes employed which is the ratio of the bandwidth at the -60-dB (from the -3-dB point) bandwidth to the -3-dB bandwidth. Shape factors of 5 and 4 to 1 are common.

2-4.5 Sweep Spectrum Analyzer

The heterodyning technique can also be used to perform an automatic spectrum analysis of a continuous wave form. The tracking filter is modified in the following manner. The input reference signal is replaced by a device called a voltage-controlled oscillator (VCO). (This device is discussed in detail in Section 6-2.3.) The VCO has a frequency output that is linearly proportional to an input voltage. If the input voltage to the VCO is a saw-tooth wave of period T, the output of the VCO varies linearly from a minimum frequency f_L to a maximum frequency f_h in a time T.

The input signal may be periodic or random. Consider for a moment the case wherein the input signal is still $\cos \omega_0 t$. As the VCO is swept from ω_L to ω_h, the output frequency, ω, of the high-pass filter following multiplier 1 varies from $\omega_L + \omega_c < \omega < \omega_h + \omega_c$. From (2-62) it is seen that $f_0(t)$ varies from

$$f_L(t) \leqq f_0(t) \leqq f_h(t)$$

where

$$f_L(t) = \frac{k_0 k_1}{2} [\cos (\omega_L + \omega_c - \omega_0)t + \cos (\omega_L + \omega_c + \omega_0)t]$$

$$f_h(t) = \frac{k_0 k_1}{2} [\cos (\omega_h + \omega_c - \omega_0)t + \cos (\omega_h + \omega_c + \omega_0)t]$$

Since the constant bandwidth filter has a center frequency ω_c, there is not any output from the narrow-band filter except when $\omega = \omega_0$. Thus the output of the meter would be zero until the swept frequency equaled that of the input signal. In the case of a random input signal there is an output of the constant bandwidth filter only for those input frequencies that are centered in the band $\omega_0 \pm \pi B$.

If the saw-tooth voltage also controls the x-axis of some readout device ($x - y$ recorder, oscilloscope—see Chapter 5) and the output of the meter circuit is the y-axis, then a plot of frequency versus amplitude is obtained. It should be mentioned that the slope of the saw-tooth signal must be such that the constant bandwidth filter has enough time to respond properly. The errors associated with sweep rate and signal bandwidth are discussed in Section 2-5. It can be noted here, however, that experience has shown that for adequate resolution of the envelope of a pulse-train spectrum, the product of the filter bandwidth (B) and pulse duration (τ) should be less than or equal to 0.1. Thus

$$B\tau \leqq 0.1$$

To guarantee adequate definition of the resulting spectrum the analyzer sweep frequency should be no greater than one-fiftieth (0.02) of the pulse period (repetition frequency). It should be realized from the results of Section 1-3.1 that a periodic pulse train is the most demanding periodic signal available. Hence these results present a conservative bound with respect to all other periodic signals.

2-4.6 Notch (Rejection) Filter

Consider one type of notch filter (called a twin-tee filter) shown in Figure 2-14a. The frequency response of such a filter is shown in Figure 2-14b. From

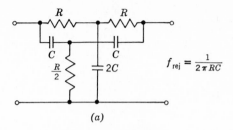

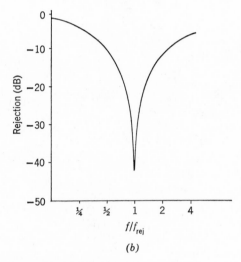

Figure 2-14 (*a*) Twin-tee filter and (*b*) its amplitude frequency response.

Figure 2-14*b* it is seen that the notch filter greatly attenuates frequencies in a band centered about f_c and permits frequencies outside this band to pass relatively unaltered.

2-4.7 Comparison of the Various Filters

Let us now compare the various filters described in Sections 2-4.2 to 2-4.4 by comparing the resolving powers of each type of filter. We consider a signal that is the sum of two different sinusoids of different frequency but of equal amplitude. The two frequencies are 940 and 1060 Hz. The results are shown in Figure 2-15 which were obtained from commercially available filters. In Figure 2-15*a* is the result of a ⅓-octave filter, Figure 2-15*b* of a 6% analyzer with an approximate selectivity of 45 dB/octave, and Figure 2-15*c* of a tracking filter with a bandwidth of 3 Hz and a −60-dB bandwidth of 12 Hz.

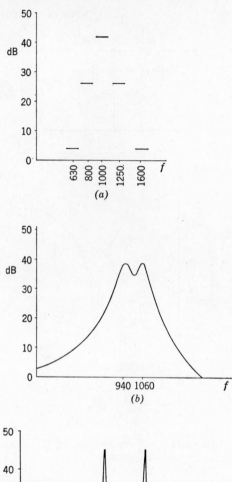

Figure 2-15 Resolution capabilities of (a) $\frac{1}{3}$-octave filter, (b) 6% analyzer and (c) a tracking filter with 3-Hz bandwidth.

2-5   Error Considerations

2-5.1   Introduction

In this section we examine the errors associated with using practical filters and band-limited systems to obtain estimates of the power spectrum for stationary random processes. Essentially we are asking the question "what statistically significant results can be obtained in a minimum period of time with a given error?"

2-5.2   Scan Rates and Averaging Time

To determine the permissible scan rates R_s, it should first be noted that two types of averaging could be used: true averaging (see Section 3-11.7) and RC averaging (recall Section 2-3.2). In the case of true averaging, the scan rate should be sufficiently slow to permit all the information in each band B_e to be reflected in each average value computation. That is, the scan rate should be less than one bandwidth per averaging time. Hence the scan-rate requirement for power spectral density measurements is

$$R_s \leqq \begin{cases} B_e/T_a = B_e{}^2/\epsilon^2 \\ B_e/4T_{RC} = B_e{}^2/4\epsilon^2 \end{cases} \tag{2-64}$$

where T_a is the true averaging time and $T_{RC} = RC$ is the RC averaging time. We have used (1-150). The factor of 4 difference between T_a and T_{RC} is to ensure that the scan rate is sufficiently slow to allow the RC filter to respond to any abrupt changes in the signal. This factor of 4 is used if the worst error, that is, the response to a step change in voltage, is to be less than 2% as compared to true averaging. If 3% error is acceptable, a factor of 2 will suffice while if 10% error is acceptable, then $T_a = T_{RC}$.

Note that there is an additional restriction on R_s required by the rise time of the bandpass filter. The rise time for an ideal bandpass filter, as shown in (2-38), is approximately $\tau_r = 1/B_e$. This means that the scan rate must be restricted to less than $B_e/\tau_r = B_e{}^2$ if the filter is to respond fully to abrupt changes in the signal. It is seen that this requirement is always less restrictive if the B_eT product is greater than unity. Since in practice it is desirable for ϵ in (1-150) to be as small as possible, this product will normally be greater than unity. Hence this additional restriction can be ignored.

If (2-64) is not adhered to, two additional errors are introduced: a decrease in the peak amplitude output of the filter and an increase or widening of B_e. Exact relations have been worked out for a filter whose $|H(f)|$ is Gaussian. That is,

$$H(f) = \exp\left[-1.3863\left(\frac{f - f_0}{B}\right)^2\right] \tag{2-65}$$

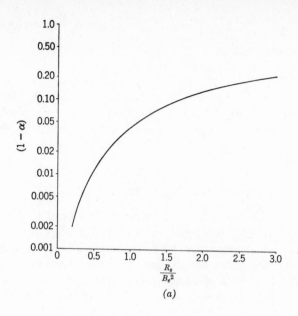

(a)

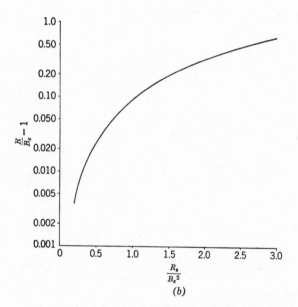

(b)

Figure 2-16 (a) Effective decrease in amplitude of a signal and (b) effective decrease in resolution capability versus R_s/B_e^2.

where B is the -3-dB bandwidth and f_0 is the center frequency. This is easily verified, for when $f = f_0$, $H(f_0) = 1.0$, and when $f = f_0 \pm B/2$, $H(f_0 \pm B/2) = 1/\sqrt{2}$. For this type of filter the maximum amplitude α is given by*

$$\alpha = C_0^{-\frac{1}{4}} \tag{2-66a}$$

where

$$C_0 = \left[1 + 0.195 \left(\frac{R_s}{B_e^2} \right)^2 \right] \tag{2-66b}$$

and $B_e \approx B$. For the analysis of random signals we use (2-64) and (1-150) to obtain

$$\alpha = (1 + 0.195\epsilon^4)^{-\frac{1}{4}} \quad \text{for true averaging} \tag{2-67a}$$

and

$$\alpha = (1 + 0.012\epsilon^4)^{-\frac{1}{4}} \quad \text{for } RC \text{ averaging} \tag{2-67b}$$

Hence for ϵ as great as 50%, $\alpha = 1.0$ for all practical purposes.

For the case of periodic (or, in general, deterministic) signals one does not require (1-150), and (2-66) is used directly. Hence the quantity R_s/B_e^2 should be kept small. Equation 2-66 is plotted in Figure 2-16a.

The relationship concerning the resolution, R, of the filter is

$$R = B_e C_0^{\frac{1}{2}} \tag{2-68}$$

where C_0 is given by (2-66b). Equation 2-68 is plotted in Figure 2-16b.

2-5.3 Total Analysis Time

To determine the total analysis time, T_s, required to analyze a given signal over a frequency range $f_r = f_H - f_L$ using either a constant bandwidth filter or a constant percentage bandwidth filter, we note that $R_s = df/dt$, where f is the frequency in hertz. For a constant bandwidth filter, (2-64) becomes

$$\frac{df}{dt} = \frac{B_e}{k_1 T} \tag{2-69}$$

and for a constant percentage bandwidth filter

$$\frac{df}{dt} = \frac{B_e}{k_1 T} = \frac{B_e^2 \epsilon^2}{k_1} = \frac{f^2 \epsilon^2 \beta^2}{k_1} \tag{2-70}$$

where $k_1 = 1$, $T = T_a$; $k_1 = 4$, $T = T_{RC}$; and $\beta = 1/Q$ is the percentage of the center frequency [see (2-58)]. Integrating (2-69) and (2-70) yields

* See ref. 17.

1. Constant Bandwidth (B_e = Constant)

$$T_s \geqq \frac{f_r k_1 T}{B_e} = \frac{f_r k_1}{B_e^2 \epsilon^2} \tag{2-71}$$

2. Constant Percentage Bandwidth (β = Constant)

$$T_s \geq \frac{f_r B_e k_1 T}{f_H f_L \beta^2} = \frac{f_r k_1}{f_H f_L \epsilon^2 \beta^2} \tag{2-72}$$

As an illustration of the above results consider the case where it is desired that the frequency resolution at 100 Hz be 5% with an accuracy of 10%. If RC averaging is used and the frequency range is from 20 to 6000 Hz, what is the total analysis time and the accuracy of the overall result for an 80% confidence level? The total analysis time for a constant bandwidth filter is $T_s = 95,680$ sec or approximately 26.5 hr, whereas for the constant percentage bandwidth analysis $T_s = 7840$ sec or approximately 2.2 hr. Using (1-151) we find that $n_e = 200$. Using Figure 1-15 we see that our result will be within approximately $\pm 11\%$ of the true mean 80% of the time.

On the other hand, if the resolution requirement was 5% at 1000 Hz, the other conditions remaining the same, and then the constant bandwidth analysis would take 957 sec while the constant percentage analysis would, of course, remain the same. However, a resolution of 5% at 1000 Hz would mean a resolution of only 50% at 100 Hz. The general conclusion, therefore, is that if resolution requirements are much higher at high frequencies than at low frequencies, then constant bandwidth analysis should be used, while if resolution requirements are the same independent of frequency, or higher at lower frequencies, then constant percentage bandwidth analysis would be the optimum analysis.

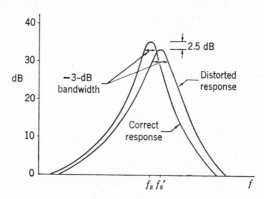

Figure 2-17 Effects of R_s/B_e^2 on amplitude sensitivity and resolution.

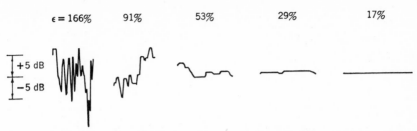

Figure 2-18 Effects of averaging time on the normalized standard error for B_e = constant.

Before leaving this section on error considerations it is worthwhile to illustrate the significance of the formulas by recording actual experimentally obtained results. The results are shown in Figures 2-17 and 2-18. Figure 2-17 shows the effects predicted by (2-66a) and (2-68). Figure 2-18 shows the effects of averaging time, for bandwidth held constant, on the normalized standard error given by (1-150).

Application 2-3   Distortion of a Sine Wave

Using a swept spectrum analyzer and an oscilloscope we can readily determine with fair accuracy the amount of distortion in a sine wave. The spectrum analyzer is swept over a frequency range which is at least 10 times the frequency of the sine wave. Recalling Section 1-3.2 we see that any measurable amplitude appearing in the oscilloscope trace occurring at integer multiples of the sine-wave frequency indicate the presence of harmonic distortion. From a rough estimate of their amplitudes relative to the amplitude of the original frequency and (1-125) an indication of the percentage harmonic distortion can be determined.

Application 2-4   Elimination of Public Address System "Squeal"

Consider a public address system comprised of a microphone, amplifier, and loudspeaker in an enclosed room (auditorium, lecture hall, etc.). In poorly designed rooms, from an acoustical viewpoint, there are reflections from the room's boundaries in certain narrow frequency bands. When the room is occupied by a large group of people, one can consider that the microphone is being subjected to a random-noise background. This noise background contains frequencies in, say, one of the frequency bands at which excessive reflections from the boundaries occur. Relative to the microphone, then, these frequencies appear amplified due to the room acoustics. The microphone in turn transmits these sounds back into the room via the loudspeaker. The loudspeaker has added more acoustic energy at these frequencies into the room so that in the next cycle a larger signal is being transmitted via the microphone-loudspeaker path. Within a matter of seconds the incoming acoustic energy exceeds the amplifier's capabilities and becomes unstable, locked onto this frequency so that a horrible, spine-tingling squeal has permeated the entire room. The same phenomenon can be caused by the speaker when his voice generates sounds containing these frequencies. To remedy this one must isolate the frequency band and insert a notch filter centered at these frequencies before the amplifier.

Application 2-5   Octave Filter and Power Spectrum

Let us re-examine the octave filter and show how under certain circumstances one can obtain the power spectral density directly from the average power readings in each octave bands. From (1-130) an estimate of the power spectral density is $G(f) = P_0/B = P_0/\beta f_c$ where P_0 is the average power in the band centered at f_c with bandwidth B. If the octave filter is a contiguous filter consisting of N frequency bands, then from (2-54) we see that the center frequency in one band is related to the center frequency in the adjacent band by

$$f_c^{(2)} = 2^n f_c^{(1)}$$
$$f_c^{(3)} = 2^n f_c^{(2)}$$
$$\vdots$$
$$f_c^{(N)} = 2^n f_c^{(N-1)}$$

where $f_c^{(j)}$ is the center frequency of the jth filter and n is the octave. Using this result we see that the spectral energy density in the jth frequency band is

$$G^{(j)}(f) = \frac{P_0^{(j)}}{2^{(j-1)n}\beta f_c^{(1)}} \qquad j = 1, 2, \ldots, N$$

The circumstance under which this relation becomes useful is when $G^{(j)}(f) = G^{(1)}(f)/2^{(j+1)n}$. This says that the power spectral density is decreasing at the rate of -3 dB/octave. This shaping is easy to obtain and is discussed in Application 6-4. Specific applications using this property are also discussed in the same chapter. If the spectral density has this property, we see that $P^{(j)} = \beta f_c^{(1)}G^{(1)}(f)$ or the power in each octave band is a constant. Thus if one is in the position to shape the input noise spectrum as stated above the effect of the increase in the bandwidth of each succeeding octave band has been cancelled, and, therefore, the magnitudes of the average power recorded in *each* octave band are divided by the same bandwidth to yield the power spectral density. Hence comparison of levels recorded in each band are immediately comparable without the formal division of this bandwidth, since we would need simply to change the amplitude scale of the entire recording to reflect this division.

2-6 Common Practical Filters

2-6.1 Introduction

In the beginning of this chapter the definition of an ideal filter is given. Since these filters are physically unrealizable, they can only be approximated by practical filters. The actual techniques to obtain a practical filter can be found in a comprehensive book on network synthesis. The purpose of this book is only to briefly describe the more common filters found in current use; namely, the Butterworth, Chebyshev, and linear-phase filters. The discussion of these filters is limited to the low-pass filter, although the extension of the discussion to bandpass and high-pass filters is straight-forward.

2-6.2 Butterworth Filters

A Butterworth filter is one which gives a maximally flat approximation to the magnitude of an ideal filter in the frequency domain. Its transfer function is of the form

$$|H(\omega)|^2 = (1 + \omega^{2n})^{-1} \qquad (2\text{-}73)$$

where ω is normalized so that at $\omega = 1$ the magnitude is down 3 dB. The integer n is called the order of the filter. A typical response is shown in Figure 2-19. Notice that for $n = 1$ (2-73) is the same as (2-46), the low-pass *RC* filter.

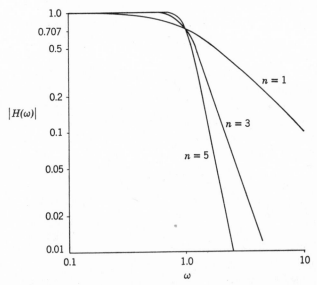

Figure 2-19 Typical transfer function of a Butterworth filter.

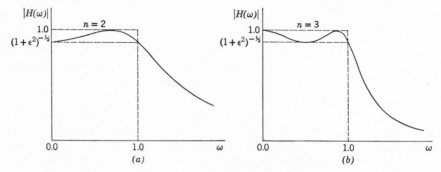

Figure 2-20   Typical transfer function of (a) second order and (b) third order Chebyshev filter.

2-6.3   Chebyshev Filters

A Chebyshev or equal ripple filter is one in which the error of approximation to the ideal-filter magnitude is evenly distributed throughout the passband in an oscillating manner. Its transfer function is given by

$$|H(\omega)|^2 = [1 + \epsilon^2 C_n^2(\omega)]^{-1} \tag{2-74}$$

where $\epsilon < 1$ is a real constant which determines the magnitude of the ripple, ω is again normalized, and $C_n(\omega)$ is the nth-order Chebyshev polynomials defined by

$$C_n(\omega) = \begin{cases} \cos{(n \cos^{-1} \omega)} & 0 \leq \omega < 1 \\ \cosh{(n \cosh^{-1} \omega)} & \omega > 1 \end{cases}$$

A typical response of second- and third-order Chebyshev filter is shown in Figure 2-20. The Chebyshev filter provides a sharper "knee" than the Butterworth filter. There is a trade-off for this sharper cutoff which appears as ripple in the magnitude response in the passband. This design is most useful when only the amplitude characteristic is important. Extensive tables for the design of these filters can be found in Ref. 14.

2-6.4   Linear-Phase Filters

In the case of the linear-phase filters amplitude response is momentarily ignored and the phase response is required to be linear with frequency; that is, $\theta(\omega) = \tau_0 \omega$. It turns out that this linear-phase response is so structured that it has what is called a maximally flat-delay response. A special case of the general linear-phase filter is the Bessel filter. There is no region in a linear-phase filter wherein the amplitude characteristic (frequency response) is a constant in the passband.

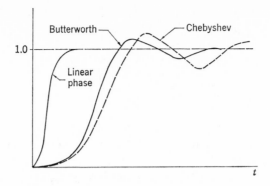

Figure 2-21 Normalized transient response of various fifth-order filters to a step input.

2-6.5 Comparison of the Three Filters

The properties of these three types of filters are best compared by examining their real-time response. The Chebyshev filter has a longer transient response as compared to the Butterworth filter and both exhibit large overshoot and undershoot, especially for higher-order filters. On the other hand, the linear-phase filters have excellent transient-response characteristics which include the minimum amount of overshoot and undershoot regardless of the order of the filter. In addition, they exhibit the fastest rise time. The results are summarized in Figure 2-21. Several cases for the time response of these filters have been given in Ref. 7. In addition, Ref. 1, 9, and 12 give numerous tables for the design of various types of filters as a function of bandwidth, selectivity, and such.

2-7 Matched Filters

In many instances one is interested in determining the presence of a pulsed signal masked by additive noise. By the presence of the signal it is meant that the fidelity of the pulse shape is not important but only whether the signal is there at some instance of time is important. The question is then asked "how is the filter designed so as to maximize the ratio of the peak value of the pulse to the rms noise?" The word pulse means a signal that has definable width, with most of its energy concentrated over a specified time interval. Expressed mathematically we want to maximize the expression

$$\frac{s_0(t)}{\sqrt{\langle n_0^2(t) \rangle_e}} \qquad (2\text{-}75)$$

where $s_0(t)$ is the signal output from a filter (without noise) and $\langle n_0^2(t) \rangle_e$ is the mean-square value of the noise output from the same filter without the signal. Equation 2-75 is the definition of the peak signal-to-rms-noise ratio. Another definition sometimes used is the ratio of the rms value of the output signal to the rms value of the output noise; that is,

$$\frac{\langle s_0^2(t) \rangle_T^{1/2}}{\langle n_0^2(t) \rangle_e^{1/2}} \tag{2-76}$$

For the results in this section we employ (2-75). The type of filter used for this purpose is called a predetection filter and is used to enhance the strength of the signal relative to noise and thereby facilitate detection. One filter designed to do just this is called a North or matched filter.

Assume for the moment that a signal that is a rectangular pulse of duration t_0 is applied to an ideal low-pass filter of cutoff frequency f_2. As was discussed in Section 2-2.5 and implied in (2-37), the peak of the output signal for small bandwidth ($f_2 \ll 1/t_0$) is small and increases with bandwidth. The rms noise in the ideal low-pass filter is proportional to $\sqrt{f_2}$ (square root of the bandwidth) and, therefore, increases at a smaller rate than the signal for small bandwidth. As the bandwidth increases to $f_2 = 1/t_0$, the signal amplitude begins to increase less rapidly with f_2. For $f_2 > 1/t_0$, the signal remains approximately at the same amplitude but with better definition of the pulse. However, the rms noise keeps increasing with increasing bandwidth (f_2) so that the ratio of peak signal to rms noise begins to decay inversely as $\sqrt{f_2}$. Thus we should expect an optimum ratio at about $f_2 = 1/t_0$. This is shown in the following analysis.

Using (2-6), (2-18), and (1-49), (2-75) can be written as

$$\frac{s_0^2(t)}{\langle n_0^2(t) \rangle_e} = \frac{\left[\int_{-\infty}^{\infty} H(f) s_i(f) e^{j2\pi f t} \, df \right]^2}{\int_{-\infty}^{\infty} |H(f)|^2 S_{nn}(f) \, df} \tag{2-77}$$

where $H(f)$ is the transfer function of the filter, $s_i(f)$ is the Fourier transform of the input signal, $S_{nn}(f)$ is the power spectral density of the input noise, and f is the frequency in hertz. Consider the following inequality (Schwartz's)

$$\left| \int_{-\infty}^{\infty} X(f) Y(f) \, df \right|^2 \leq \int_{-\infty}^{\infty} |X(f)|^2 \, df \int_{-\infty}^{\infty} |Y(f)|^2 \, df \tag{2-78}$$

If we consider the input noise to be white noise, then $S_{nn}(f) = N_0/2$, and substitution of (2-78) into (2-77) yields

$$\frac{s_0^2(t)}{\langle n_0^2(t) \rangle_e} \leq \frac{2}{N_0} \int_{-\infty}^{\infty} |s_i(f)|^2 \, df = \frac{2E}{N_0} = d^2 \tag{2-79}$$

wherein we have used (1-55) and E is the total energy in the signal. The quantity d is called the peak signal-to-noise ratio. For any arbitrary signal (pulse) of duration t_0 and filter the signal-to-noise must be less than ideal; that is,

$$\frac{s_0^2(t)}{\langle n_0^2(t)\rangle_e}\frac{1}{d^2} < 1 \tag{2-80}$$

From (2-78), however, it is apparent that the equality sign holds only when $X(f) = Y^*(f)$; that is, when they are complex conjugates. Therefore, to obtain the maximum peak signal-to-noise ratio, which changes the inequality sign of (2-80) to an equal sign, the filter response must be of the form

$$H(\omega) = [ks_i(\omega)e^{j\omega t_0}]^* = ks_i^*(\omega)e^{-j\omega t_0} \tag{2-81}$$

A filter whose response is given by (2-81) is said to be a *matched filter*. Thus except for a possible amplitude and delay factor of the form $ke^{-j\omega t_0}$, the transfer function of a matched filter is the complex conjugate of the spectrum of the signal to which it is matched. For this reason a matched filter is often called a "conjugate" filter.

Using (2-2b), (2-81), and the fact that $f(-\omega) = f^*(\omega)$, the impulse response $K(\tau)$ becomes

$$K(\tau) = \frac{k}{2\pi}\int_{-\infty}^{\infty} s_i^*(\omega)e^{j\omega(\tau - t_0)}\,d\omega$$

or

$$K(\tau) = ks_i(t_0 - \tau) \tag{2-82}$$

Using (2-81) it is seen that the output of the filter is

$$s_o(t) = \frac{1}{2\pi}\int_{-\infty}^{\infty} |s_i(\omega)|^2\, e^{j\omega(t - t_0)}\,d\omega \tag{2-83}$$

In particular $s_o(t)$ will have a maximum at $t = t_0$; thus

$$s_o(t_0)\big|_{\max} = \frac{1}{2\pi}\int_{-\infty}^{\infty} |s_i(\omega)|^2\,d\omega = E \tag{2-84}$$

and, therefore, the peak value of $s_o(t)$ is *independent* of the choice of t_0. This time of occurrence can be varied by simply adjusting the phase characteristics, the $e^{-j\omega t_0}$ term, of the matched filter.

Detection by use of a matched filter is equivalent to determining the autocorrelation function of the input signal. Using (2-82) in (2-1) yields,

$$s_o(t - t_0) = \int_{-\infty}^{\infty} s_i(\lambda)s_i(\lambda + [t - t_0])\,d\lambda \tag{2-85}$$

This is just the form for the autocorrelation function of a signal shifted an amount in time t_0. Notice that $s_o(t - t_0)$ is a maximum when $t = t_0$ as previously pointed out.

The peak signal-to-noise ratio will now be calculated for two types of filters when a rectangular pulse of the type given by (1-56) is applied to the filter. The results are normalized with respect to the matched filter. Thus dividing (2-77) by E and letting $S_{nn}(f) = N_0/2$ the following quantity will compare any filter to a matched filter when the noise is white noise:

$$\gamma = \frac{\left\{ \int_{-\infty}^{\infty} H(\omega)s_i(\omega)e^{j\omega t_0}\, d\omega \right\}^2}{\int_{-\infty}^{\infty} |s_i(\omega)|^2\, d\omega \int_{-\infty}^{\infty} |H(\omega)|^2\, d\omega} \tag{2-86}$$

It is easily seen from (2-78) and (2-81) that for the case of a matched filter $\gamma = 1$, which is its maximum value. From (1-57) the frequency spectrum of the signal is

$$s_i(\omega) = \frac{\sin (\omega\tau/2)}{\omega\tau/2} \tag{2-87}$$

where τ is the pulse duration and the quantity $V\tau$ has been set equal to unity for convenience. Thus

$$\int_{-\infty}^{\infty} |s_i(\omega)|^2\, d\omega = \frac{2}{\tau} \int_{-\infty}^{\infty} \frac{\sin^2 x}{x^2}\, dx = \frac{2\pi}{\tau} \tag{2-88}$$

Now consider the following filters.

1. Ideal Low-Pass Filter

Using (2-11), (2-87), and (2-88), (2-86) becomes

$$\gamma = \frac{2}{\pi a} [Si(a)]^2 \tag{2-89}$$

where $a = \omega_c\tau/2$ and $Si(a)$ is the sine integral given by (2-36). Using a table of the sine integral it is found that γ has a maximum at $a = 2.2$. Therefore, $f_c\tau = 2.2/\pi = 0.7$. At this bandwidth γ is found to be 0.83 as compared to 1 for the optimum filter, thus corresponding to a relative deterioration of 0.8 dB. The result given by (2-89) is plotted in Figure 2-22 as a function of $f_c\tau$. Although the maximum ratio is at $f_c\tau = 0.7$, it is seen that the maximum is very broad and varies less than 1 dB from approximately $f_c\tau = 0.4$ to $f_c\tau = 1.0$. Recall that the rise time for the filter is $\tau_r = 0.5/f_c$.

2. Low-Pass RC Filter

Using (2-45), (2-87), and (2-88), (2-86) becomes

$$\gamma = \frac{(1 - e^{-2a})^2}{a} \tag{2-90}$$

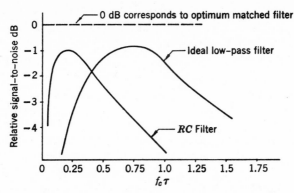

Figure 2-22 Peak signal-to-rms-noise for various filters compared to a matched filter.

Plotting this result in Figure 2-22 shows that the amplitude maximum occurs at $f_c\tau = \tau/(2\pi RC) = 0.2$. At this bandwidth the filter output is only 1 dB worse than the matched filter. For $f_c\tau = 0.5$, γ is 2.3 dB worse than the matched filter so that the variation with bandwidth is again relatively small.

2-8   Optimum Prediction (Interpolation) Filter

Consider the case wherein a signal $s(t)$ in a background of additive noise $n(t)$ is passed through a filter whose output is $z(t)$. It is desired that the signal $s(t)$ be separated from $n(t)$ by means of a linear filter having a transfer function $H(\omega)$ such that the mean-square difference $\langle[z(t) - s(t)]^2\rangle_T$ between the desired and actual outputs is minimized. If $E(t)$ is the mean-square error, it can be shown* that the filter transfer function is given by

$$H(\omega) = \frac{S_s(\omega)e^{-i\omega\alpha}}{S_s(\omega) + S_n(\omega)} \qquad (2\text{-}91)$$

where $S_s(\omega)$ and $S_n(\omega)$ are the power spectra of the signal and the noise, respectively, and α is the prediction time interval. With this filter the theoretical minimum mean-square error is given by

$$\langle E^2(t)\rangle_T \Big|_{\min} = \int_0^\infty \frac{S_s(\omega)S_n(\omega)}{S_s(\omega) + S_n(\omega)}\,d\omega \qquad (2\text{-}92)$$

which is valid for all values of t. In general the filter given by (2-92) is not realizable physically, but it can be approximated physically within arbitrarily

* See ref. 1 of Chapter 1, pp. 153 ff.

close tolerances by allowing a sufficiently long delay α. This solution, however, does have a practical value in that it specifies a near-optimum filter for situations in which a long but finite delay is of no consequence. Further, (2-92) has the intuitively satisfying interpretation that the best smoothing filter results when the input signal-to-noise spectrum is more heavily weighted where the signal-to-noise spectral density ratio is large, that is, the spectral regions which contain largely signal power are treated favorably and those containing mostly noise power unfavorably.

Notice the difference between this type of filter and a matched filter. In matched filters the signal shape is of no consequence only its peak value, at some time t_0, above the noise. In the optimum filter one is interested in obtaining the best possible recovery of the signal shape from the noise.

2-9 Summary

2-9.1 General

The impulse response function of a filter is its time-varying output signal when its input signal is a very short duration pulse. The Fourier transform of the impulse function is the transfer function of the filter which gives the amplitude and phase characteristics of the filter as a function of frequency. The Fourier transform of the output signal is equal to the product of the Fourier transform of the input signal and the transfer function of the filter. The spectral density of the output signal from a filter is equal to the product of the spectral density of the input signal and the square of the absolute value of the transfer function of the filter. From the application of a harmonic signal to a filter one can completely determine its amplitude and phase characteristics of its transfer function. The cross-correlation of the input signal to the filter with its output signal when the input signal is broad-band white noise yields the impulse-response function of the filter.

2-9.2 Ideal Filters

An ideal filter is defined as one in which the amplitude of its transfer function is a constant with frequency and its phase angle is proportional to frequency. This constant of proportionality, t_d, is the time it takes the impulse response function to reach its maximum. In addition t_d is the delay time of the input signal through the filter. The bandwidth of an ideal filter is determined by its cutoff frequencies.

2-9.3 Basic Properties of Filters

The transient response of a filter is a function of the envelope of the input signal. The quantity t_d is the time it takes the output signal to reach approximately half its final or steady-state value. The rise time of a filter is specified by the time it takes the output signal to rise from 10 to 90% of its final value.

The product of the rise time of the filter and its bandwidth is a constant. This constant ranges from 0.3 to 1.0. To transmit a pulse relatively undistorted through a filter, the filter's bandwidth must be much greater than the reciprocal of the duration of the pulse, or equivalently, the duration of the pulse must be much longer than the rise time of the filter.

2-9.4   Practical Filters

General.   A practical filter is defined by its −3-dB frequencies, selectivity, and amplitude variations of its transfer function within its passband. The amplitude variations are specified within some tolerance around a mean value. The frequencies at which the amplitude frequency response decreases to $1/\sqrt{2}$ of the mean value in the passband are defined as the cutoff frequencies or the −3-dB frequencies of the filter. The rate of attenuation of the amplitude of the transfer function from the cutoff frequencies defines the selectivity of the filter. The effective bandwidth of a practical filter is a measure of the closeness of the filter to an ideal filter with regard to the total energy transfer through the filter. Very narrow constant bandwidth filters are described by their shape factors which are the amount of attenuation per doubling of filter bandwidth.

Classification.   Filter types are defined by the relationship of their center frequency to their cutoff frequencies. The center frequency is defined as the geometric mean between the cutoff frequencies. A constant percentage bandwidth filter is one in which the bandwidth is a percentage of the center frequency. A constant bandwidth filter is one in which the bandwidth and the cutoff frequencies define the center frequency. Variable bandwidth filters are ones in which the cutoff frequencies vary. The ratio of the filter center frequency to its bandwidth is called the quality or Q of the filter. When the Q of a filter is high (>10) the geometric mean and the arithmetic mean, which can be used to obtain the center frequency, are for all practical purposes the same.

To minimize errors when using filters to perform spectral analysis on random signals the product of the effective bandwidth of the filter and the averaging (integrating) time should be as large as possible to reduce both the normalized statistical error and overall measurement error for a given confidence level. In addition, for either random or deterministic signals, the ratio of the scan rate to the square of the equivalent bandwidth should be less than 0.2 for negligible loss in amplitude sensitivity and decrease in filter resolution.

Common Types of Practical Filters.   Three of the more common types of filters are the Butterworth, Chebyshev, and linear-phase filters. The Butterworth filter has a maximally flat amplitude response in its passband. The Chebyshev filter has an amplitude response which varies within a specified amount in its passband. The attenuation of frequencies in the neighborhood of the -3-dB frequencies and outside of the passband is much greater for a Chebyshev filter than for a Butterworth filter. The amplitude response of a linear-phase filter is nowhere a constant in its passband. The transient-response characteristics are the best for the linear-phase filter and the worst for the Chebyshev filter.

The simplest filters are the low-pass and high-pass *RC* filters. The product *RC* is called the time constant of the filter. The cutoff frequencies for both of these filters are the same and are proportional to the reciprocal of the time constant. When used at frequencies well above its cutoff frequency a low-pass *RC* filter acts as an integrator while when used at frequencies well below its cutoff frequency a high-pass *RC* filter acts as a differentiator.

Matched Filter.   A matched filter is one type of predetection filter which is used to enhance the peak signal-to-noise ratio. The peak signal-to-noise ratio is the ratio of the peak value of the signal to the rms value of the noise without the signal. The transfer function of a matched filter is the complex conjugate of the Fourier transform of the input signal. The matching of the filter to the signal in this manner is equivalent to performing the auto-correlation on the input signal plus noise.

References

1. E. Christian and E. Eisenmann, *Filter Design Tables*, John Wiley and Sons, Inc., New York, 1967.
2. M. S. Ghausi, *Principles and Design of Linear Active Circuits*, McGraw-Hill Book Co., New York, 1965.
3. E. H. Guillemin, *Communication Networks*, Vol. II, John Wiley & Sons, Inc., New York, 1935.
4. E. H. Guillemin, *Synthesis of Passive Networks*, John Wiley & Sons, Inc., New York, 1957.
5. E. H. Guillemin, *Theory of Linear Physical Systems*, John Wiley & Sons, Inc., New York, 1963.
6. C. W. Helstrom, *Statistical Theory of Signal Detection*, Pergamon Press, New York, 1960.
7. K. W. Henderson and W. H. Kantz, "Transient Response of Conventional Filters," *IRE Trans. Circuit Theory*, **CT-5**, No. 4 (December 1958), p. 333.
8. M. Javid and E. Brenner, *Analysis, Transmission, and Filtering of Signals*, McGraw-Hill Book Company, Inc., New York, 1963.
9. J. H. Mole, *Filter Design Data for Communication Engineers*, John Wiley and Sons, Inc., New York, 1952.

10. M. Schwartz, *Information, Transmission, Modulation, and Noise,* McGraw-Hill Book Company, Inc., New York, 1959.

11. L. W. Sepmeyer, "Bandwidth Error of Symmetrical Bandpass Filters Used for Analysis of Noise and Vibration," *J. Acoust. Soc. Amer.,* **34,** No. 10 (October 1962), pp. 1653–1657.

12. J. G. Truxal, Ed., *Control Engineers Handbook,* McGraw-Hill Book Co., New York, 1958.

13. G. L. Turin, "An Introduction to Matched Filters," *IRE Trans. Information Theory,* **IT-6,** No. 3 (June 1960), pp. 311–329.

14. L. Weinberg, "Network Design by Use of Modern Synthesis Techniques and Tables," *Proc. Nat Electron. Conf., Chicago,* **12,** 1956, p. 794.

15. A. I. Zverev, *Handbook of Filter Synthesis,* John Wiley and Sons, Inc., New York, 1967.

16. "Spectrum Analysis," Hewlett-Packard Co., Palo Alto, Calif., Application Note 63, August 1968.

17. "Spectrum Analyzer Techniques Handbook," 4th ed., Polarad Electronics Corp., Long Island City, N.Y., 1962.

3

AMPLIFIERS

3-1 Introduction

In many measurements it is necessary to increase the voltage or power output of some signal. This may be necessary because some element in a measurement system provides too small an output for other devices in the system. Increasing the magnitude or power of a signal requires amplification. Amplifiers have, in general, two basic functions; one is to amplify signals that are too low in level for their intended application, and the second is to isolate circuits from other circuits. In its ideal state an amplifier draws no power from the system to which it is connected, does not alter the form of the input signal in any way, has an instantaneous response-time, has constant gain in a given bandwidth, and can be used to control any other system without affecting its output signal.

Amplifiers can be broadly classified as voltage amplifiers or power amplifiers. The purpose of a voltage amplifier is to increase the magnitude of some input voltage. Thus the ideal voltage amplifier has the transfer function

$$E_{\text{out}}(\omega) = A \, E_{\text{in}}(\omega)$$

where $E_{\text{out}}(\omega)$ is the output voltage of the amplifier, $E_{\text{in}}(\omega)$ is the input voltage to the amplifier, and A is a constant, independent of the frequency; that is, the amplifier transfer function is independent of ω.

The purpose of the power amplifier is to control some device that is connected to its output. Thus the ideal power amplifier has the transfer function

$$P_{\text{out}}(\omega) = A_0 \, P_{\text{in}}(\omega)$$

where $P_{\text{out}}(\omega)$ is the output power of the amplifier, $P_{\text{in}}(\omega)$ is the input power to the amplifier, and A_0 is a constant, independent of the frequency.

Since practical amplifiers cannot perform to the ideal standards, this chapter introduces definitions of meaningful quantities that are used to indicate an amplifier's performance with respect to the ideal case, and further, discusses some standard techniques that can be used to overcome the inherent limitations of an amplifier. Numerous amplifier configurations that are useful in a wide variety of applications are also discussed.

97

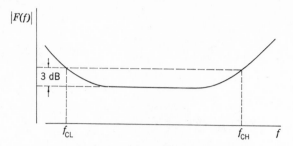

Figure 3-1 Typical noise spectrum of an amplifier.

Before proceeding it should be pointed out that the definitions which were introduced in Section 2-2.1 for filters can be applied directly to amplifiers.

3-2 Noise Characterization

3-2.1 Introduction

In this chapter noise is defined as any spurious or undesired disturbance that tends to obscure or mask the input signal to the amplifier. There are three broad classifications of noise sources found in an amplifier: flicker (or $1/f$) noise, thermal noise, and shot noise. Flicker noise begins to influence noise in an amplifier at some relatively low frequency, f_{CL}, the low-frequency noise corner and is inversely proportional to frequency at a 3 dB/octave rate (see Figure 3-1). As yet, flicker noise is not completely predictable mathematically. However, in transistor amplifiers, the flicker noise corner can be lowered to some extent by transistor fabrication techniques, and is usually of minor importance except for very weak signals at very low frequencies.

Neither thermal noise nor shot noise is frequency dependent, and both exhibit uniform noise output (behave as band-limited white noise) through the entire useful frequency range of the amplifier. However, the internal gain of the amplifier components does vary with frequency and decreases as frequency increases. For constant gain at high frequencies the gain of the amplifier components must be increased. The increase in gain at high frequency also amplifies the noise. The 3-dB point of increasing noise is called the upper high-frequency noise corner f_{CH}.

3-2.2 Thermal Noise

Thermal noise is generated by all conductors not at absolute zero temperature. The noise arises from the fact that the thermal agitation in the conductors give rise to electrical energy. The thermal power generated depends only on the temperature and the bandwidth over which it is measured. This

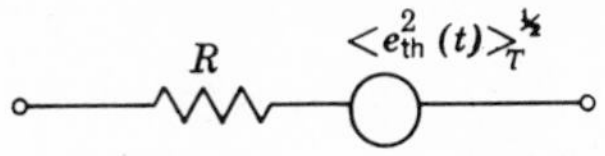

$$R \qquad \langle e_{\text{th}}^2(t) \rangle_T^{\frac{1}{2}}$$

Figure 3-2 Equivalent circuit of a noisy resistor.

agitation is so completely random that when terminals are attached to the medium, it is found that the voltage at the terminals covers all frequencies. The mean-squared value of the thermal noise $\langle e_{\text{th}}^2(t) \rangle_T$ may be represented by a resistor in series with a noise generator and is described by

$$\langle e_{\text{th}}^2(t) \rangle_T = 4kTR\,\Delta f \qquad (3\text{-}1)$$

where k is Boltzmann's constant (1.38×10^{-23} joules/°K), T is the temperature in degrees Kelvin, Δf is the bandwidth in hertz, and R is the resistance in ohms. As shown in Figure 3-2 a noisy resistor has been equated to a noiseless resistor plus a noise voltage generator.

3-2.3 Shot Noise

A second type of noise is termed shot noise. Shot noise occurs whenever a current flows. The mean-squared noise current is described mathematically by

$$\langle i_{\text{sh}}^2(t) \rangle_T = 2qI_{\text{dc}}\,\Delta f \qquad (3\text{-}2)$$

where q is the charge of an electron (1.6×10^{-19} coulombs), I_{dc} is the dc current flow in amperes, and Δf is the bandwidth in hertz.

The equivalent circuit for shot noise is that of a current generator in parallel with a noiseless conductance g_0, where g_0 is the effective conductance of the region through which the current flows (see Figure 3-3a). For transistors and semiconductor diodes the conductance g_0 is given by $g_0 = qI_{\text{dc}}/kT$, where q, I_{dc}, k, and T have been previously defined. It should be noted that a similar expression exists for vacuum tubes and can be found in ref. 1. Using the expression for the conductance of the diode, it is possible to convert the mean-squared shot-noise current generator to an equivalent voltage generator. The expression for the equivalent voltage generator shown in Figure 3-3b is given mathematically by

$$\langle e_{\text{sh}}^2(t) \rangle_T = 2kTr_0\,\Delta f \qquad (3\text{-}3)$$

where $r_0 = 1/g_0$.

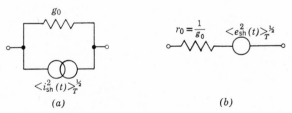

(a)   (b)

Figure 3-3 Equivalent current (a) and voltage (b) circuits for a noisy diode.

3-2.4   Optimum Noise Source

It has been shown that the lowest noise from a transistor depends not only on the equivalent resistance but also on the amount of current flowing. A transistor has three main internal noise generators: shot noise in the emitter-base, junction thermal noise in the base resistance, and shot noise in the collector base junction. A complete discussion of optimum noise source is beyond the scope of this book; however, it has been shown* that the optimum noise-source resistance is directly proportional to the square of the sum of the emitter and base resistance, and inversely proportional to the ratio of the collector current and emitter current, the square of the transistor frequency-dependent operating parameters, and the current gain. Thus the optimum resistance is not necessarily the one with the lowest possible resistance.

3-3   Noise in an Amplifier

For noise considerations, any amplifier may be considered a linear two-port network characterized by a noise-voltage generator in series with the input. Figure 3-4 shows a noiseless amplifier with an input resistance, R_a, and a noise generator with its generator resistance, R_g. All the noise in the amplifier (flicker, shot noise, and thermal noise) is represented by the noise generator. The internal resistance of the noise generator is represented by R_g; that is, in Figure 3-4 the noise generator itself has no resistance. It is assumed that all of the noise-generator voltage e_g will appear across the amplifier's input. It is also assumed that the rms output voltage of the amplifier can be divided by the gain to give the magnitude of the equivalent input noise-voltage generator. This method is useful when a comparison between amplifiers is to be made and the operational bandwidth of the amplifiers is known. Another method which considers source resistance and bandwidth is discussed subsequently.

Consider the case where the gain of the amplifier is unity and the noise-generator input is zero. The rms output of the amplifier is, from (3-1),

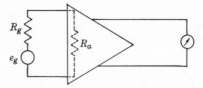

Figure 3-4   Noiseless amplifier connected to a noise generator.

* See ref. 9, for example.

$\langle e_{\text{th}}^2(t)\rangle_T^{1/2}$. However, the noise contributed by the amplifier will always be greater than that given by (3-1). Thus (3-1) denotes the minimum theoretical noise from an amplifier. An equation found useful in expressing the amount of noise generated in an amplifier is the noise figure which relates the minimum noise obtainable from an amplifier (due to R_g) to the actual noise from the amplifier. The noise figure, NF, is given by

$$\text{NF} = \frac{d_i^2}{d_o^2} \tag{3-4}$$

where d_i^2 and d_o^2 are the input and output signal-to-noise power ratios, respectively, given by the square of (2-76). If the device contained no internal sources of noise, the entire available output voltage would come from the input termination and the noise figure would be unity. Additional noise sources within the amplifier cause the noise figure to exceed unity. Excess of the noise figure over unity thus represents the amount of degradation from the ideal amplifier. The noise figures are frequently given in decibels, and since the ratios d_i^2 and d_o^2 are in power, the conversion is simply $10 \log_{10}$ (NF).

The actual calculation of noise voltages in amplifiers requires the combination of noise voltages due to shot noise and thermal noise since all amplifiers have active and passive elements. Total noise voltages may be found simply by summing mean-squared voltages from the shot-noise generator and the thermal-noise generator [recall (1-138) with $\tau = 0$].

The total noise in an amplifier is represented by an equivalent noise generator at the input. If an amplifier has a gain of 60 dB (1000) and it is found that the output noise is 100 mV, the equivalent noise generator would have a value of 100 μV. Thus if the equivalent-noise-generator rms value is known, the output noise can be found for any amplifier gain. Such a method of specifying the equivalent noise is termed "referred to the input."

It is apparent that the real amplifier is different from the ideal amplifier since the real amplifier will contribute some additional noise. This noise, at best, can approach the theoretical limit expressed by (3-1). This additional noise will also result from the supply power to the amplifier and differences in ground potential due to connections in the system as described in Appendix C. When the input-signal rms value approaches the rms value of the equivalent noise generator, the input signal will be lost in the noise. Note that if the signal is filtered, the signal can again be recovered from the noise, but the bandwidth of the amplifier and consequently the noise figure has, in effect, been decreased.

As discussed in Section 3-2.4 the optimum noise figure is not necessarily associated with the lowest input resistance. In order to completely specify the noise characteristics of any amplifier, the noise at different frequencies and

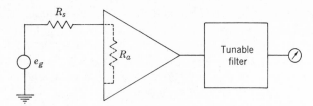

Figure 3-5   Functional diagram to obtain noise characteristics of an amplifier.

source resistances must be measured. To accomplish this measurement a test circuit as shown in Figure 3-5 is used.

A white-noise generator with a calibrated attenuator is adjusted for zero output from the amplifier and a source resistance R_s is inserted between the noise generator and the amplifier. As explained in Section 3-6, the equivalent output resistance of the noise generator must be much less than the lowest source resistance to be investigated. The filter is adjusted to the desired center frequency and the rms output of the filter is recorded. The noise-generator output is then increased from zero until the output of the filter is $\sqrt{2}$ times its former value. This corresponds to a doubling of output power. The output of the noise generator is thus the total noise due to the amplifier plus the source resistance. By varying the source resistance while maintaining the center frequency of the filter constant the total noise as a function of source resistance can be determined. By varying the center frequency while maintaining the source resistance constant, total noise as a function of frequency can be determined. The noise figure is then calculated for each source resistance and center frequency using (3-1) and (3-4) where R is the source resistance and Δf is the effective bandwidth of the filter. Contours of constant noise figure are then plotted as a function of the log of the frequency versus the log of the source resistance. A typical set of noise-figure contours are shown in Figure 3-6.

Noise-figure contours are very useful. When the source resistance and operating frequency are fixed by experimental limitations, the contours of several amplifiers are used to determine the amplifier with the minimum noise. If an amplifier is already available, the experiment may be designed such that the source resistance and frequency are chosen for the lowest noise figure point on the contour. An amplifier cannot usefully amplify signals which are below the internal noise level. Since noise-figure contours provide complete information on internal amplifier noise, they can be used to determine the minimum signal that can be detected using a particular amplifier.

Now consider the case where two amplifiers are cascaded; that is, the output signal from one amplifier is fed into another amplifier immediately following

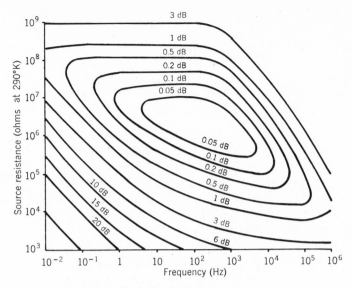

Figure 3-6   Typical noise-figure contours.

as shown in Figure 3-7. This is the situation where two stages of amplification are used in an amplifier or one amplifier does not provide enough gain and a second amplifier is connected to the system. The individual amplifiers have noise figures NF_1 and NF_2 and gains A_1 and A_2, respectively. It is not too difficult to show that the total noise figure NF_T for the cascaded amplifiers is

$$NF_T = NF_1 + \frac{NF_2 - 1}{A_1} \tag{3-5}$$

It is clear that the relative contribution of two cascaded amplifiers is most dependent on the last amplifier. Thus if the gain of the first amplifier, A_1, is large ($A_1 \gg NF_2 - 1$), the second network contributes negligibly to the overall noise figure. Thus the overall noise figure remains at NF_1. Hence for the

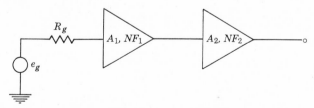

Figure 3-7   Cascaded amplifiers with different gains and noise figures.

best signal-to-noise ratio for two amplifiers in cascade, the first amplifier should be used at maximum gain and the second amplifier should be used at minimum gain (both amplifiers having the same order of magnitude noise figure).

Equation 3-5 can be generalized to the case of N cascaded amplifiers to give

$$NF_T = NF_1 + \frac{NF_2 - 1}{A_1} + \cdots + \frac{NF_N - 1}{A_1 A_2 \cdots A_{N-1}} \qquad (3\text{-}6)$$

where A_N is the gain and NF_N is the noise figure for the Nth amplifier, respectively.

It is clear from the preceding discussion that there will always be some noise present in any amplified signal. Consider now the errors introduced by noise in a signal. The total mean-squared value of a signal is determined from (1-138) with $\tau = 0$. Thus

$$\langle e_m{}^2 \rangle_e = \langle e_s{}^2 \rangle_e + \langle e_n{}^2 \rangle_e \qquad (3\text{-}7)$$

where $e_m(t)$ is the measured signal, $e_s(t)$ is the desired signal, and $e_n(t)$ is the noise. The percentage error due to the noise is defined as

$$\text{error} = \left[1 - \left(\frac{d_0{}^2}{1 + d_0{}^2} \right)^{1/2} \right] \times 100\% \qquad (3\text{-}8)$$

where d_0 is given by (2-76). The percentage error versus signal-to-noise ratio, d_0, is plotted in Figure 3-8.

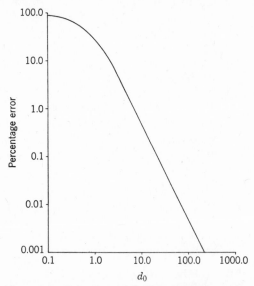

Figure 3-8 Percentage error versus signal-to-noise ratio.

3-4 Linearity

It was stated that the gain of an ideal amplifier is not only independent of frequency but was also independent of the amplitude of its input. It is clear that the output voltage of an amplifier cannot exceed the external voltage supplying the amplifier in the case of a voltage amplification or the external power supplying the amplifier in the case of a power amplification. Thus for a practical amplifier, the gain or the input to the amplifier must be given some limit. Hence

$$E_{out}(\omega) = AE_{in}(\omega) \qquad |E_{in}| < \frac{k_v}{A}$$

$$P_{out}(\omega) = A_0 P_{in}(\omega) \qquad |P_{in}| < \frac{k_p}{A_0}$$

where E and P are the voltage and power, respectively.

When the input parameter to the amplifier is increased such that the associated output parameter exceeds k_v and k_p, the voltage or power cannot be increased. If the input is increased beyond this limit the output will be "clipped" (i.e., limited) as shown in Figure 3-9. Consider the case where the

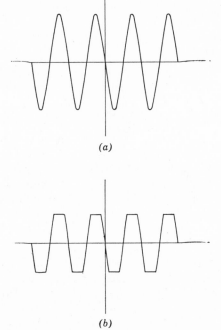

(a)

(b)

Figure 3-9 Input (a) and output (b) signal of an amplifier operating in its nonlinear region.

input parameter is a sinusoidal signal. Suppose the input is increased such that the output parameter would be $10k_v$ if the output was not limited. Because of the limiting action of the amplifier the output would approach a square wave with its fundamental frequency equal to that of the sinusoidal signal. It should be noted that other types of distortion can occur besides those due to limiting. In general, clipping consists mainly of odd harmonics being generated, while other types of distortion can generate even harmonics. The generated harmonics are the result of the amplifier being operated in its nonlinear amplitude region. Such harmonic generation is termed harmonic distortion and has already been discussed in Section 1-3.2.

When an amplifier is operated in its nonlinear region, a signal is generated that was not present in the input signal. The upper linear amplitude level determines the maximum output signal that can be obtained from the amplifier and still meet the requirements of the ideal amplifier; this level is stated as some percentage harmonic distortion. That is, all the harmonics generated by distortion are some specified percentage of the amplitude of the fundamental frequency. Notice the parallel of linearity of an amplifier with that given for a stationary filter in Section 2-2.1.

3-5   Dynamic Range

It has been shown that all amplifiers have a lower limiting signal level. If an input signal below this level is amplified, the signal will be indistinguishable from the inherent noise of the amplifier. It has also been shown that all amplifiers have a higher limiting parameter. If an input parameter is greater than this level, harmonic distortion will result. The region between the lower limit and the upper limit is termed the dynamic range of the amplifier. Since the dynamic range of an amplifier is, in general, large, it is usually expressed in decibels. It is desirable to have an amplifier with as small a noise level as possible. This noise level will depend on the physical parameters of the amplifier. A voltage amplifier may have a dynamic range of 60 dB. If the amplifier had a noise level of 10 μV, the maximum voltage that could be applied to the input without excessive harmonic distortion would be 10 mV. Thus the amplier could only be used for signals between 10 μV to 10 mV. To increase the range of voltages (or power) which an amplifier may amplify, an attenuator (or potentiometer) usually precedes the actual amplifier. The purpose of the attenuator is to reduce the magnitude of the input signal such that it is within the dynamic range of the amplifier. It should be noted that the attenuator has not increased the dynamic range of the amplifier but has increased only the range of voltages with which the amplifier may be used.

With an amplifier it is necessary that a provision be made to connect (couple) the input signal to the first stage of amplification. Sometimes a single element will provide the needed amplification for a signal, but more often a number of stages are required. The amplification is performed in stages, with the individual stages connected in cascade so that the output of one stage feeds into the input of the next. The coupling of the signal, either to the first amplifier or between stages, may be performed directly, capacitively, or inductively.

If the signal passed through the amplifier contains frequencies that essentially go to zero (dc), the stages are direct coupled. Direct coupling means that all of the elements are (ideally) without energy storage or memory; thus the elements have the same transfer function at all frequencies down to zero. Clearly resistors are used in this case. Since the frequency response of the amplifier goes down to dc, any dc component of the signal will limit the useful range of the amplifier. Also any slow change caused by aging or temperature change will appear as additive low-frequency noise to the input signal and will be amplified along with the input signal. By introducing a capacitor in series with the input, thus forming a high-pass filter, the low-frequency signals are blocked.

Transformer coupling also blocks the dc signal and drift and allows only ac signals to be amplified. In addition the transformer provides the added advantage of impedance matching, which is discussed in the next section. However a transformer with wide frequency response is relatively difficult to make and is comparatively large. For these reasons, transformer coupling is generally used only when the load has a relatively low impedance.

3-6 Impedance Matching

The introduction of any measuring instrument into a measured medium always results in the extraction of some energy from the medium. This extraction of energy changes the value of the measured quantity from its undisturbed state. Consider, for example, the amplifier shown in Figure 3-10. When the amplifier is connected to the output of a voltage generator, the circuit is changed and the value of the generator voltage, e_g, is no longer

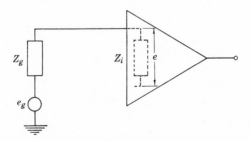

Figure 3-10 A voltage generator connected to an amplifier.

the same as when the circuit was not loaded. It is an easy matter to show that the voltage across Z_i is related to e_g by

$$e = e_g\left[1 + \frac{Z_g}{Z_i}\right]^{-1} \tag{3-9}$$

where Z_g and Z_i are, in general, complex quantities. Let us assume that

$$\frac{Z_g}{Z_i} = \frac{a + jb}{c + jd} \tag{3-10}$$

Then it can be easily shown that

$$\left|\frac{e}{e_g}\right|^2 = \frac{(c^2 + d^2)}{(a + c)^2 + (b + d)^2} \tag{3-11}$$

Using (3-11) we now define the percentage error caused by a finite input impedance Z_i of the amplifier as

$$\% \text{ error} = \left(1 - \left|\frac{e}{e_g}\right|\right) \times 100\% \tag{3-12}$$

Let us now examine the percentage error for several combinations of Z_g and Z_i.

1. Z_g and Z_i Resistive. In this case $Z_g = R_g$, $Z_i = R_i$, and from (3-10) $b = d = 0$ and $a/c = R_g/R_i$. Thus (3-11) and (3-12) give, respectively,

$$\left|\frac{e}{e_g}\right| = \left(1 + \frac{R_g}{R_i}\right)^{-1} \tag{3-13a}$$

$$\% \text{ error} = \frac{R_g}{R_i}\left(1 + \frac{R_g}{R_i}\right)^{-1} \times 100\% \tag{3-13b}$$

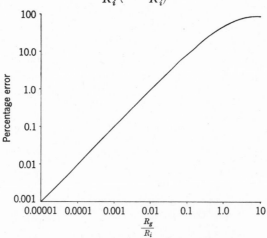

Figure 3-11 Percentage error as a function of resistive loading.

Equation (3-13) shows very clearly that R_g/R_i must be equal to 0.01 for the error to be less than 1%. Equation 3-13b is plotted in Figure 3-11.

2. Z_g Resistive, Z_i Complex. Let $Z_g = R_g$ and $Z_i = R_i/(1 + j\omega\tau_i)$, where $\tau_i = R_iC_i$. Z_i therefore, represents the case where the input resistance of the amplifier, R_i, is in parallel with a capacitor, C_i. This capacitance is called the shunt capacitance. From (3-10) we see that $a = \alpha$, $b = \omega\alpha\tau_i$, $c = 1$, $d = 0$, and $\alpha = R_g/R_i$. Then (3-11) and (3-12) become, respectively,

$$\left|\frac{e}{e_g}\right| = [(1 + \alpha)^2 + \alpha^2\omega^2\tau_i^2]^{-\frac{1}{2}} \tag{3-14a}$$

$$\% \text{ error} = \{1 - [(1 + \alpha)^2 + \alpha^2\omega^2\tau_i^2]^{-\frac{1}{2}}\} \times 100\% \tag{3-14b}$$

From the form of (3-14a) we see that this combination of impedances behaves like a low-pass filter. In this regard recall (2-46). Equation 3-14a, b shows that in order to have a very high cutoff frequency with the smallest error $R_g/R_i \ll 1$ and $\alpha\tau_i = R_gC_i \ll 1$. This last inequality requires that C_i be as small as possible. Equation 3-14b is plotted in Figure 3-12.

3. Z_g Capacitive, Z_i Complex. In this case $Z_g = 1/(j\omega C_g)$ and Z_i is that given in the preceding example. From (3-10), $a = 1$, $b = \omega\tau_i$, $c = 0$,

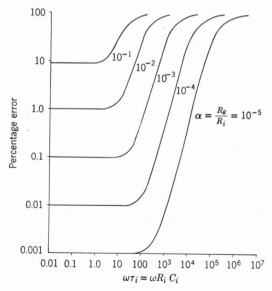

Figure 3-12 Percentage error as a function of $\omega\tau_i$ for various α.

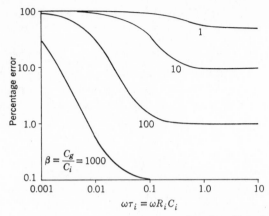

Figure 3-13 Percentage error as a function of $\omega\tau_i$ for various β.

$d = \omega R_i C_g$, and (3-11) and (3-12) become, respectively,

$$\left|\frac{e}{e_g}\right| = \beta\omega\tau_i[1 + \omega^2\tau_i^2(1 + \beta)^2]^{-\frac{1}{2}} \tag{3-15a}$$

$$\% \text{ error} = \{1 - \beta\omega\tau_i[1 + \omega^2\tau_i^2(1 + \beta)^2]^{-\frac{1}{2}}\} \times 100\% \tag{3-15b}$$

where $\beta = C_g/C_i$. Equation 3-15b is plotted in Figure 3-13. From Figure 3-13 we see that for minimum error at low frequencies C_g/C_i should be as large as possible. From (3-15a) we see that this combination of impedances behaves as a high-pass filter [recall (2-51)]. Equation 3-15a, b shows that in order to have a very low cutoff frequency with the smallest error $R_i C_i \gg 1$. However, from (3-15a) we see that it is not good to have C_i large so that C_g/C_i decreases, for the amplitude of the signal input is attenuated. Thus R_i must have a large value such that the above inequality is satisfied.

Although in the last two examples the capacitance C_i was denoted the shunt capacitance, it need not have been restricted to this. Other capacitances in the actual case would also be included in C_i besides the shunt capacitance; namely, stray capacitance, C_s, and the capacitance of the connecting cables, C_c. These capacitances appear in parallel with C_i and, therefore, the symbol C_i in (3-14a, b) and (3-15a, b) can be replaced by $C_T = C_i + C_s + C_c$. From the preceding discussion, therefore, one sees the necessity of keeping C_T as small as possible.

The above discussion is also applicable to attenuators. Consider the simple resistor attenuator shown in Figure 3-14. From (3-9) and (3-10) we see that when $C_1 = C_2 = 0$, the attenuation, a_T, is

$$a_T = \frac{R_2}{R_1 + R_2} \tag{3-16}$$

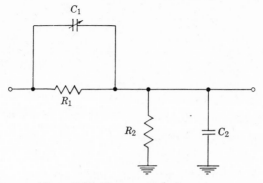

Figure 3-14 Compensated attenuator.

which would be independent of frequency if it were not for the inevitable stray capacitance C_2 that shunts R_2. The capacitance is present even though shielded cable is used to isolate the input lead from stray fields such as those of the power line (see Appendix C). The capacitance may also be the input capacitance of the input stage of an amplifier. The resistors R_1 and R_2 are large enough so that the input impedance of the attenuator does not load the input signal source. The attenuator may be compensated so that its attenuation is independent of the frequency by shunting R_1 with an adjustable capacitance C_1 as shown in Figure 3-14. In this case we see that $Z_g/Z_i = \alpha(1 + j\omega\tau_2)/(1 + j\omega\tau_1)$ where $\alpha = R_1/R_2$, $\tau_1 = R_1 C_1$, $\tau_2 = R_2 C_2$, and, therefore, $a = \alpha$, $b = \alpha\omega\tau_2$, $c = 1$, and $d = \omega\tau_1$. From (3-11) we see that in order to make Z_g/Z_i independent of frequency we must have $b = qd$ and $a = qc$, where q is an arbitrary constant. Solving for q gives $a = bc/d$. Substituting in the above values for a, b, c, and d yields $C_1 = R_2 C_2/R_1$.

Suppose now that a maximum power transfer is required to a load on the output of the amplifier. In this case the output impedance of the amplifier and the input impedance of the load on the amplifier are considered in their general form as complex quantities. The output impedance of the amplifier is $Z_0 = R_0 + jX_0$ and the impedance of the load is $Z_L = R_L + jX_L$. The active power to the load is given by

$$P = \frac{E^2 R_L}{|Z_L + Z_0|^2} = \frac{E^2 R_L}{(R_L + R_0)^2 + (X_0 + X_L)^2} \qquad (3\text{-}17)$$

where E is the rms magnitude of the voltage. To maximize (3-17) we see that with R_L constant the greatest value will occur when $X_L = -X_0$. With $X_L = -X_0$, (3-17) becomes

$$P = \frac{E^2 R_L}{(R_L + R_0)^2} \qquad (3\text{-}18)$$

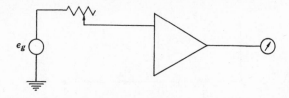

Figure 3-15 Method for determining the equivalent input resistance of an amplifier.

Differentiating (3-18) with respect to R_L and equating the result to zero yields $R_L = R_0$. Thus maximum power occurs when $Z_L = Z_0^*$, where the asterisk denotes the complex conjugate. Such a load receives half the power output of the source, the other half going into the resistance R_0.

Application 3-1 Measurement of Equivalent Input Resistance

Equation 3-13a suggests a simple method for measuring the equivalent input resistance of an amplifier. A signal generator, that is, oscillator or random noise generator, with equivalent output resistance much less than that of the expected equivalent input resistance of the amplifier under test is connected in series with a variable resistance to the amplifier as shown in Figure 3-15. A voltmeter is connected to the output of the amplifier; the variable resistor is set to its minimum value; and the signal generator is adjusted to some convenient output voltage. Then the variable resistance is adjusted until the output voltage of the amplifier indicates one half the previously read output voltage. The value of the variable resistor is measured giving the value of the equivalent input resistance of the amplifier.

Application 3-2 Impedance Matching of Capacitance Transducers

A capacitive transducer is one in which mechanical to electrical conversion is affected by mechanically induced changes in capacitance. The capacitance between two conducting plates separated by an insulator is inversely proportional to the distance between the plates (provided that the plates are large compared to their separation so that edge effects can be neglected).

The most common method used with capacitance transducers is to apply a large dc voltage, called a "polarization" voltage, from a high resistance source to one plate. This places a constant charge, q, on the capacitor. One plate is held fixed while motion of the second plate causes a change in the capacitance. The voltage V across a capacitor is related to the charge on it by

$$V = \frac{q}{c} \approx \frac{q}{d}$$

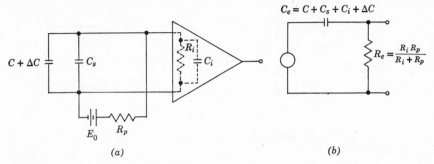

Figure 3-16 (*a*) Polarization circuit for a capacitance transducer and (*b*) its equivalent circuit.

provided that the distance d between the plate is much smaller than the smallest dimension of the plates.

Consider the dc polarization circuit and amplifier shown in Figure 3-16a. The polarization voltage E_0 is connected to the capacitance transducer through a power supply resistance R_p. The capacitances C_s, C_i, and the resistance R_i represent stray capacitance between transducer and output device, the input capacitance, and input resistance of the amplifier, respectively.

The circuit shown in Figure 3-16a can be reduced to the equivalent circuit shown in Figure 3-16b. The equivalent circuit shown in Figure 3-16b is that of a high-pass filter (see Figure 2-12). Recall that the cutoff frequency is given by $f_c = 1/(2\pi R_e C_e)$. Thus for the transducer to provide a low-frequency output either C_e or R_e must be large. However, note that the sensitivity of the capacitance transducer is proportional to $\Delta C/C$. Hence for low-frequency response R_e must be made large. This large value for R_e is usually obtained using a cathode or emitter follower, which is simply a noninverting amplifier in an appropriate configuration. See Section 3-11.3. The high-frequency cutoff of capacitance transducers depends upon the physical parameters of the transducer. Examples of capacitance transducers are condenser microphones and dynamic capacitance displacement gauges.

Application 3-3   Impedance Matching of Capacitance-Like Transducers

A piezoelectric material is a nonconducting crystal that when mechanically stressed generates an electrical charge. The piezoelectric element acts like a capacitor, and thus the charge variations appear as voltage variations across a suitable high resistance. The equivalent circuit of a piezoelectric transducer is shown in Figure 3-17a, where C_a is the equivalent capacitance of the transducer, C is the capacitance of the cable, and R_i and C_i are the input resistance and capacitance of the amplifier, respectively. The equivalent

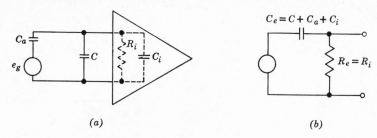

(a)   (b)

Figure 3-17 (a) Piezoelectric transducer connected to an amplifier and (b) its equivalent circuit.

circuit shown in Figure 3-17a can be reduced to the circuit shown in Figure 3-17b. Thus the low-frequency cutoff is again given by $f_c = 1/(2\pi R_e C_e)$. Hence the piezoelectric transducer is, in general, used with a high-impedance amplifier when a low-cutoff frequency is desired.

We see that the low-frequency response of a capacitance transducer is limited by the impedance of the amplifier. However, its high-frequency response is limited by the physical parameters of the transducer itself. A capacitance transducer is a lightly damped (high Q) system and is unusable near its resonant frequencies. Examples of piezoelectric transducers are accelerometers, hydrophones, and crystal microphones.

Application 3-4   Impedance Matching of Inductive Transducers

When an electrical conductor is moved through a magnetic field, a voltage, e, is induced in the conductor which is proportional to the strength of the magnetic field, the length of the exposed conductor, and the velocity of motion, v. Thus $e = kv$, where k is a constant that depends upon the strength of the magnetic field and the length of the conductor. Inductive transducers are reversible transducers. That is, if a voltage is applied to the transducer, a velocity output results. Examples of inductive transducers are electrodynamic exciters and loudspeakers.

In the case where the inductive transducer is used as an electrodynamic exciter (often called a vibration exciter) a sinusoidal signal is generated, amplified, and applied to a coil which is suspended between a static magnetic field. The suspension is such that it is rigid to torsion but relatively compliant to flexure. A test object is mounted on a table attached to the coil and the response as a function of frequency is measured. In this case it is necessary for maximum power to be transferred. Thus the output impedance of the amplifier is designed so that its output resistance is equal to the resistive part of the vibration-exciter input impedance. However, the imaginary component of the impedance will vary greatly with frequency. For this reason a variable capacitance bank is sometimes added in series with the predominantly inductive load to tune out the reactive part of the impedance, thus getting maximum power transfer.

3-7 Frequency Response and Rise Time of Cascaded Amplifiers

Consider a capacitive coupled amplifier with a transfer function of

$$H(\omega) = \frac{A_0}{1 + j\omega\tau}$$

for a single stage [see (2-45)]. Since the gains of noninteracting cascaded stages are multiplicative, the transfer function for n identical stages is

$$H(\omega) = \frac{A_0^n}{(1 + j\omega\tau)^n}$$

If the input to the n cascaded amplifiers is a suddenly impressed and maintained voltage of magnitude E_0, it can be shown that the output, $e_0(t)$, of the nth amplifier is

$$\frac{e_0(t)}{E_0 A_0^n} = \left[\frac{\alpha^n}{n!} + \frac{\alpha^{n+1}}{(n+1)!} + \cdots \right] e^{-\alpha} \tag{3-19}$$

where $\alpha = t/\tau$. Notice that as n increases there is introduced a progressively larger delay and that the rise time increases with n. If the rise time τ_r is defined as the amount of time required for the signal to go from 0.1 to 0.9 of the final value it can be found from (3-19) that the rise time of two identical stages is 1.5 times that of a single stage. Values for all additional identical stages are given in Table 3-1.

Table 3-1 Rise-Time Ratios for Cascaded Amplifiers

n	2	3	4	5	6	7	8	9	10
τ_{r_n}/τ_{r_1}	1.5	1.9	2.2	2.5	2.8	3.0	3.3	3.45	3.6

Using (2-39) we see that

$$\tau_{r_n} B_n = \tau_{r_1} B_1 \tag{3-20}$$

where B_n is the overall bandwidth of the n stage amplifier and τ_{r_n}/τ_{r_1} is given in Table 3-1. Thus if one wanted the overall bandwidth of an n-stage amplifier to be B_n, then (3-20) gives the value of the bandwidth of each individual stage, B_1. For example, a two-stage amplifier with an overall response of 1 MHz would require that each individual amplifier have a response of 1.5 MHz.

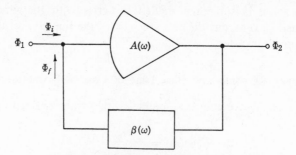

Figure 3-18 Single-loop feedback system.

An approximate (within 10%) time-domain counterpart to (3-20) is

$$\tau_r = 1.1 \sqrt{\tau_{ro} + \sum_{i=1}^{n} \tau_{r_i}^{2}} \qquad (3\text{-}21)$$

where τ_{r_i} is the rise time of each stage and τ_{ro} is the overall rise time of the input signal. This formula is valid for amplifiers wherein the overshoot in each stage to a suddenly impressed voltage is less than 5%.

3-8 Negative Feedback in Amplifiers

Negative feedback in amplifiers is often used to obtain significant improvements in the performance of the amplifier. When feedback is used the resulting modification depends upon external elements, which are usually passive, and not on the amplifier itself. The price paid for obtaining these improvements is mainly the introduction of possible instability and reduced gain in the system.

Before proceeding further consider the block diagram for an amplifier with feedback shown in Figure 3-18. The feedback loop is defined as a closed path, consisting of the forward path and the feedback path. If the feedback loop is open so that no signal is returned, that is, $\Phi_f = 0$, the open-loop transfer function or the forward amplifier gain function, $A(\omega)$, is defined as

$$A(\omega) = \frac{\Phi_2}{\Phi_i} \qquad (3\text{-}22)$$

where Φ is a current or voltage quantity. The feedback function $\beta(\omega)$ is defined as

$$\beta(\omega) = \frac{\Phi_f}{\Phi_2} \qquad (3\text{-}23)$$

In the idealized model, none of the input signal is transmitted through the $\beta(\omega)$ path and none of the output signal is transmitted in the reverse direction through the amplifier. From Figure 3-18 we see that

$$\Phi_1 = \Phi_i - \Phi_f \qquad (3\text{-}24)$$

From (3-22), (3-23), and (3-24), the basic feedback equation is obtained. This equation is the network function or the gain function of the closed-loop system and is given by

$$a(\omega) = \frac{\Phi_2}{\Phi_1} = \frac{A(\omega)}{1 - A(\omega)\beta(\omega)} \qquad (3\text{-}25)$$

where $A(\omega)$ is positive for voltage feedback (termed series feedback) or $A(\omega)$ is negative for current feedback (termed shunt feedback). Note that the application of feedback has modified the open-loop transfer function by a factor $[1 - A(\omega)\beta(\omega)]^{-1}$. If the magnitude of $1 - A(\omega)\beta(\omega)$ is greater than unity, negative feedback occurs. Negative feedback is also called degenerative feedback.

In amplifiers with no feedback, the overall gain $a(\omega)$ varies considerably with changes in the gain of the individual tubes or transistors. The use of negative feedback reduces the sensitivity of the amplifier to such changes. Sensitivity is a measure of the degree of dependence of one quantity upon the value of another quantity. Quantitatively, the sensitivity of the overall gain $a(\omega)$ with respect to a given parameter k is defined as

$$S = \frac{\Delta a/a}{\Delta k/k} \qquad (3\text{-}26)$$

The above definition may be interpreted as a normalized percentage change in $a(\omega)$ with respect to a percentage change in k. An ideal system, according to the definition in (3-26), has zero sensitivity.

From (3-25) and (3-26)

$$S = \frac{\Delta a/a}{\Delta A/A} = \frac{1}{1 - A\beta} = \frac{1}{F} \qquad (3\text{-}27a)$$

or

$$\frac{\Delta a}{a} = \frac{1}{F}\frac{\Delta A}{A} \qquad (3\text{-}27b)$$

Thus any change in sensitivity has been reduced by a factor of $1/F$. Note further that if $|A(\omega)\beta(\omega)| \gg 1$, (3-25) reduces to

$$a(\omega) \approx -\frac{1}{\beta(\omega)}$$

Since $\beta(\omega)$ is usually a passive network, $a(\omega)$ is almost independent of the active elements.

The distortion due to the nonlinearity of the active device and the disturbance due to noise and the power-supply ripple voltage are reduced by

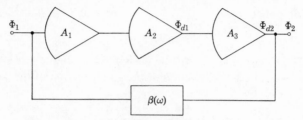

Figure 3-19   A single-loop feedback of a 3-stage amplifier subjected to extraneous signals.

the use of negative feedback. Note, however, that if the disturbance or noise occurs at the input stage, feedback has no effect. Consider the block diagram of a single-loop feedback amplifier shown in Figure 3-19, where Φ_{d1} and Φ_{d2} are the extraneous signals, such as distortion, which are assumed to be represented by disturbing signals. For convenience, the amplifier is assumed to be composed of three internal amplifying stages.

The output signal Φ_2 in terms of the input signal Φ_1 and the disturbing signals Φ_{d1}, and Φ_{d2} is

$$\Phi_2 = \frac{A_1 A_2 A_3 \Phi_1 + A_3 \Phi_{d1} + \Phi_{d2}}{1 - A_1 A_2 A_3 \beta} \tag{3-28a}$$

For $|A_1 A_2 A_3 \beta| \gg 1$, (3-28a) reduces to

$$\Phi_2 = -\frac{\Phi_1}{\beta} - \frac{\Phi_{d1}}{A_1 A_2 \beta} - \frac{\Phi_{d2}}{A_1 A_2 A_3 \beta} \tag{3-28b}$$

Notice that the effect of Φ_{d2} at the output has been reduced by a factor of $1/(A_1 A_2 A_3 \beta)$, whereas that of Φ_{d1} is reduced by the lesser amount $1/(A_1 A_2 \beta)$. Thus the magnitude of the reduction depends to a large extent upon the origin of the disturbing signal. Since nonlinear distortion increases with the level of the output signal, the assumption that the distortion signal is generated in the final stage of the cascaded amplifier is reasonable. The noise signal at the input of the first stage, however, is most critical since feedback is no help in improving the signal-to-noise ratio.

3-9   Operational Amplifiers

An operational amplifier is a high-gain, direct-coupled amplifier with high-input impedance and low-output impedance and which uses feedback for control of its response characteristics. All operational amplifiers provide access to an inverting input, referred to as the minus (−) grid or minus (−) input. A positive-going signal connected at this input produces a negative going signal at the output. Thus if there is no feedback connection, the amplifier shifts the input signal $180°$ and amplifies it some amount say $A(\omega)$. This gain and phase shift is called the open-loop gain and is usually denoted $-A(\omega)$. Some operational amplifiers provide access to a noninverting input, referred to as the positive $(+)$ grid or positive $(+)$ input. A positive-going signal connected at this point produces a positive-going signal at the output. Throughout this section if only one input is shown it is assumed to be the inverting input.

The capabilities and limitations of operational amplifiers are defined by simple equations and rules which are based on a set of criteria that an operational amplifier must meet. Effective use of these relationships requires knowledge of the conditions under which each is applicable so that errors resulting from approximations commonly used are held to a minimum.

The following sections discuss primarily the dynamic properties of the amplifier rather than its dc properties. Techniques are given for minimizing the amount of dc voltage at the output, termed dc offset, since this limits the dynamic range of the operational amplifier. The dc characteristics are included when appropriate.

3-10 Basic Properties of Operational Amplifiers

3-10.1 Introduction

The equivalent circuit for an operational amplifier is shown in Figure 3-20. The quantity Z_{io} represents the equivalent input impedance of the amplifier and Z_o represents the output impedance. Typical ranges of values for these and other commonly used operational amplifier parameters are given in Table 3-2. The open-loop voltage gain of the operational amplifier is denoted by $A(\omega)$. The equivalent input impedance is specified for two cases: differential

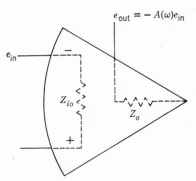

Figure 3-20 Equivalent circuit of an operational amplifier.

impedance and common-mode impedance. The consequences of these two specifications will become obvious in subsequent discussions.

Open-loop gain is defined as the ratio of the output voltage e_{out} to the input voltage e_{in}. Gain is specified at dc by the manufacturer and a plot of voltage gain versus frequency is usually included. The gain may also be given in terms of slewing rate, usually expressed in volts per microsecond. Slewing rate defines the maximum rate of change of output voltage for a sudden change in the input voltage. For a harmonic signal with amplitude V_m a first approximation of slewing rate, S, is given by

$$S = 2\pi f V_m \times 10^{-6} \quad \text{V}/\mu\text{sec} \tag{3-29}$$

The large-signal and small-signal response characteristics of operational amplifiers differ substantially due to dynamic nonlinearities or transient saturation. An amplifier will not respond to large-signal changes as rapidly as the small-signal bandwidth characteristics would predict because of the slewing rate limitation. Full power response is the maximum frequency measured at unity closed-loop gain for which rated output voltage can be obtained for a sinusoidal signal at rated load without some specified percentage harmonic distortion.

3-10.2 Inverting Feedback Configuration

An operational amplifier operated with an inverting feedback configuration is shown in Figure 3-21. The load resistor R_L is assumed to be large so that its effect on the transfer characteristics is negligible. The effects of a finite R_L were discussed in Section 3-6. Most operational amplifiers require a significant flow of current at both inputs. For this condition, the dc paths to ground for

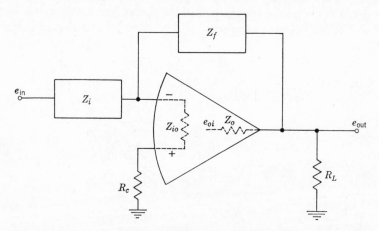

Figure 3-21 An operational amplifier in an inverting feedback configuration.

each input must be equal for a minimum dc offset voltage at the output. Dc offset is the amount of dc voltage, referred to the input, which will appear at the output superimposed on the amplified signal. If the offset voltage is large, the dynamic range of the amplifier will be limited. Usually, an external potentiometer is used to bring any inherent dc offset to zero. The actual equivalent configuration will become clear when the summing configuration is discussed. The value and particular configuration is usually specified by the amplifier manufacturer. In addition, it is assumed that

$$\frac{1}{R_c} = \left(\frac{1}{Z_i} + \frac{1}{Z_f + Z_o}\right)_{\omega=0} \tag{3-30}$$

For very large load resistance ($R_L \to \infty$) the closed-loop gain G_{CLI} is given by

$$G_{\text{CLI}} = \frac{e_{\text{out}}}{e_{\text{in}}} = \frac{-\dfrac{Z_f}{Z_i} + \dfrac{1}{A(\omega)\beta'}}{1 + \dfrac{1}{A(\omega)}\left(\dfrac{1}{\beta} + \dfrac{1}{\beta'}\right)} \tag{3-31}$$

where

$$\frac{1}{\beta'} = \frac{Z_o}{Z_i}\left(1 + \frac{R_c}{Z_{io}}\right) \tag{3-32a}$$

and

$$\frac{1}{\beta} = 1 + \frac{Z_f + Z_o + R_c}{Z_{io}} \tag{3-32b}$$

If

$$\frac{1}{A(\omega)}\left(\frac{1}{\beta} + \frac{1}{\beta'}\right) \ll 1 \tag{3-33}$$

then (3-31) reduces to

$$G_{\text{CLI}} \approx -\frac{Z_f}{Z_i} \tag{3-34}$$

The input impedance of the operational amplifier is given by

$$Z_{\text{in}} = Z_i\left[1 + \frac{1}{A(\omega)\beta_1}\left(1 + \frac{1}{A(\omega)\beta_2}\right)^{-1}\right] \tag{3-35}$$

where

$$\frac{1}{\beta_1} = \frac{Z_f}{\beta' Z_o}\left(1 + \frac{Z_o}{Z_f}\right) \tag{3-36a}$$

$$\frac{1}{\beta_2} = \frac{1}{\beta} + \frac{1}{\beta'} - \frac{Z_o}{Z_i} \tag{3-36b}$$

If $1/A(\omega)\beta_1 \ll 1$ and $1/A(\omega)\beta_2 \ll 1$, then (3-35) reduces to

$$Z_{\text{in}} \approx Z_i \qquad (3\text{-}37)$$

We see that when the gain of the operational amplifier is large compared to the feedback impedance, then the input impedance is Z_i. This implies the existence of a condition known as a virtual ground at the summing point. That is, the summing point is at ground potential even though there is no electrical connection between this point and ground. Furthermore, no current flows into the negative terminal of the amplifier when the open-loop gain is large.

The closed-loop output impedance is the ratio of the output voltage to the output current and can be written as

$$Z_{\text{out}} = -G_{\text{CLI}} \frac{Z_o}{A(\omega)\beta_3} \qquad (3\text{-}38\text{a})$$

where

$$\frac{1}{\beta_3} = \frac{Z_i}{Z_{io}} + \left(1 + \frac{Z_i}{Z_f}\right)\left(1 + \frac{R_c}{Z_{io}}\right) \qquad (3\text{-}38\text{b})$$

and G_{CLI} is given by (3-31). If we employ the same assumption used to arrive at (3-34), namely (3-33), then (3-38a) simplifies to

$$Z_{\text{out}} \approx \frac{Z_o}{A(\omega)\beta_3'} \qquad (3\text{-}39\text{a})$$

where

$$\frac{1}{\beta_3'} = \frac{Z_f}{Z_{io}} + \left(1 + \frac{Z_f}{Z_i}\right)\left(1 + \frac{R_c}{Z_{io}}\right) \qquad (3\text{-}39\text{b})$$

Thus (3-39a) shows that the output impedance of an operational amplifier in an inverting configuration is reduced by the open-loop gain times β_3'.

3-10.3 Noninverting Feedback Configuration

The general circuit for an operational amplifier in a noninverting configuration is shown in Figure 3-22. Equations for the transfer function and closed-loop input and output impedances are given in this section. These equations assume that the resistance of the load is large enough to be neglected. It is also assumed that the dc return paths to ground give a minimum dc offset.

It can be shown that the closed-loop gain function for the noninverting amplifier is given by

$$G_{\text{CLN}} = \frac{e_{\text{out}}}{e_{\text{in}}} = \frac{(1 + Z_f/Z_i) + 1/(A(\omega)\beta_4)}{1 + 1/(A(\omega)\beta_4')} \qquad (3\text{-}40)$$

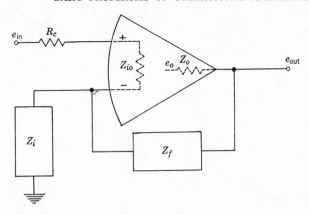

Figure 3-22 An operational amplifier in a noninverting configuration.

where

$$\frac{1}{\beta_4} = \frac{Z_o}{Z_{io}} \tag{3-41a}$$

$$\frac{1}{\beta_4'} = \frac{1}{\beta} + \frac{1}{\beta'}\left(1 + \frac{Z_f}{Z_o}\right) \tag{3-41b}$$

If $1/(A(\omega)\beta_4) \ll 1$ and $1/(A(\omega)\beta_4') \ll 1$, then (3-40) becomes

$$G_{\text{CLN}}' \approx 1 + \frac{Z_f}{Z_i} \tag{3-42}$$

The input impedance of the noninverting amplifier is given by

$$Z_{\text{in}} = Z_{io}A(\omega)\frac{1 + 1/(A(\omega)\beta_4)}{1 + \dfrac{Z_f}{Z_i} + \dfrac{Z_o}{Z_i}} \tag{3-43}$$

If we employ the same approximation used to obtain (3-42) and further, that $Z_o/Z_i \ll 1$, (3-43) simplifies to

$$Z_{\text{in}} \approx \frac{A(\omega)Z_{io}}{G_{\text{CLN}}'} \tag{3-44}$$

where G_{CLN}' is given by (3-42). We see from (3-44) that the input impedance of a noninverting operational amplifier is equal to the product of the intrinsic input impedance times the ratio of the open-loop gain to the closed-loop gain.

As in the inverting configuration, the closed-loop output impedance for the noninverting configuration is the ratio of the output voltage to the output current. It can be shown that the output impedance is

$$Z_{out} = \frac{Z_o G_{CLN}}{A(\omega)\beta_5} \qquad (3\text{-}45)$$

where

$$\frac{1}{\beta_5} = \frac{Z_f/Z_{io} + G'_{CLN}(1 + R_c/Z_{io})}{G'_{CLN}} \qquad (3\text{-}46)$$

and G_{CLN} is given by (3-40) and G'_{CLN} by (3-42). Using the assumptions that were used to arrive at (3-42), (3-45) becomes

$$Z_{out} \approx \frac{Z_o}{A(\omega)\beta'_3} \qquad (3\text{-}47)$$

where β'_3 is given by (3-39b). We note that (3-47) is identical with (3-39a) for the inverting configuration.

3-10.4 Stability of Operational Amplifiers

Recalling the closed-loop gain response for the inverting and noninverting operational amplifier configurations given by (3-31) and (3-40), respectively, we see that their denominators are of the form $1 + A(\omega)\beta$. It can be shown that if the phase angle of the feedback term $A(\omega)\beta$ is 180° while the magnitude of the term is still unity or greater, oscillations will occur. If controlled, these oscillations are useful as a signal generator as discussed in Chapter 6. If the term is greater than unity, the oscillations will build up until limiting

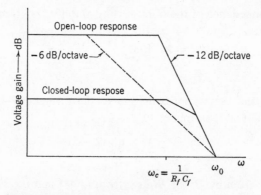

Figure 3-23 Open- and closed-loop frequency response with a stabilizing feedback capacitor.

occurs. This limiting decreases the gain, and thus the entire feedback term, until unity magnitude at a 180° phase angle is achieved.

For an unconditionally stable configuration the open-loop response, $A(\omega)$, must intersect the closed-loop response with a slope that is less than 12 dB/octave. The advantage of using an almost 12 dB/octave rolloff to the $A(\omega)$ response is that it provides more loop gain at higher frequencies. A way of implementing this increase in bandwidth is to add a feedback capacitor, C_f, in parallel to the feedback resistor, R_f, so that at the cutoff frequency, $f_c = 1/(2\pi C_f R_f)$, the closed-loop response is rolling off at 6 dB/octave. So long as the *rate* of closure between the open-loop response and closed-loop response is less than 12 dB/octave, the closed-loop response will be stable. This is illustrated in Figure 3-23.

3-10.5   Maximum Input and Output Voltage Characteristics

In Section 3-4 linearity of an amplifier was discussed. The same discussion holds for an operational amplifier. However, since the amplifier is modified with feedback, certain additional information is also required. The maximum input voltage refers to the maximum voltage allowable at the input. Recall, however, that the inverting amplifier is operating at a virtual ground and hence the input impedance is Z_i, the impedance of the external component. Therefore, this information is only needed for the noninverting amplifier and special circuits such as comparison circuits, which will be discussed in a subsequent section. For differential input circuits, that is, one wherein the combination of the inverting and noninverting configurations are used, the circuit may be operated at a voltage greater than virtual ground. In such cases the common-mode voltage is the maximum amount of voltage that can be applied without damage to the amplifier. The common-mode voltage is the difference between the feedback voltage and the input voltage to the positive input. Notice that this definition is meaningful only for the noninverting configuration or a variation thereof.

The output voltage and the output current are the maximum amount of voltage and current that can be obtained from the operational amplifier. Note that the current must be sufficient to drive any subsequent load and the feedback components. One problem not apparent from an amplifier's current specification is that of capacitive loads. A capacitive load may not only exceed the output current capabilities of the amplifier at high frequency, it may also cause the amplifier to become unstable. The amount of current drawn by a capacitor is

$$i_L = C_L \frac{dv}{dt} \tag{3-48}$$

where i_L is the current drawn by the load, C_L is the capacitive load, and dv/dt is the rate of change of voltage (in this case the slewing rate). Thus if an amplifier has a maximum current of 50 mA, and a slewing rate of 1000 V/ μsec, the maximum total capacitive load that could be driven would be 50 pF. Any increase in capacitance would put the amplifier into current limiting, causing an effective phase shift and an unstable condition.

When the output or input voltage rating of an operational amplifier is exceeded, it is possible, in some designs, for circuitry to become saturated. That is, it takes a finite amount of time for the amplifier to return to a normal state *after* the input voltage is removed. Such time is called the overload recovery time which may be greater than the period of the highest frequency at which the amplifier may be used. Thus unlike a normal amplifier, the output will remain at some fixed value and the input will again drive the amplifier into saturation. Recall that with a normal amplifier the result would be harmonic distortion. However, the output voltage of the operational amplifier will hold at some fixed dc value, usually the value of one of the voltages supplying the amplifier. Conditions of saturation are called latching.

3-10.6   Dc Characteristics of Operational Amplifiers

With the exception of dc offset, the dc characteristics of operational amplifiers are of minor importance in dynamic applications. Offset voltage is the amount of dc voltage superimposed on the output signal. If the offset voltage is large, the dynamic range of the amplifier will be limited. The first step in minimizing dc offset voltage is to include the equal paths to ground as explained in Sections 3-10.2 and 3-10.3. If R_c [recall (3-30)] is included, the offset voltage will be small. Additionally, external potentiometers can be used to set the offset to zero. This operation is performed, in general, with the inputs connected to ground. Other dc characteristics are discussed where appropriate in the particular configuration under consideration.

3-11   Operational Amplifiers Used for Signal Conditioning

3-11.1   Introduction

The following sections contain some commonly used linear applications of operational amplifiers. Nonlinear applications are considered in Section 3-12. In this regard the reader is referred to refs. 13 and 20 where more than 250 applications, both linear and nonlinear, are presented in reasonable detail. The inverting amplifier or the noninverting amplifier are considered basic to all other linear circuits. The offset errors given are static errors and should be minimized for maximum dynamic range. Typical values for operational amplifier properties are given in Table 3-2.

Table 3-2   Comparative Specifications of Currently Available Operational Amplifiers

Specifications	General-Purpose Differential	Chopper-Stabilized	FET-Input	Integrated Circuit
Dc voltage gain, open-loop	10^4–10^6	10^6–10^9	10^4–10^6	10^2–10^6
Gain rolloff	6 dB/octave	6 dB/octave	6 dB/octave	—
Slewing rate limit	0.1–1000 V/μs	0.1–250 V/μs	0.1–250 V/μs	0.1–100 V/μs
Initial input				
Offset voltage	0.1–1 mV	10–100 μV	0.5–5 mV	0.5–10 mV
Bias current	1–500 nA	10–200 pA	0.5–100 pA	2–1000 nA
Input impedance, open-loop				
Differential	10^5–10^7 ohms	10^5–10^6 ohms	10^{10}–10^{12} ohms	10^4–10^8 ohms
Common mode	10^7–10^9 ohms	—	10^{10}–10^{12} ohms	—
Input voltage (maximum)				
Differential	5–15 V	0.5–15 V	5–15 V	1–15 V
Common mode	5–20 V	—	8–11 V	1–15 V
Common mode rejection	50–120 dB	—	54–120 dB	60–100 dB
Output voltage	10–100 V	10–150 V	10–100 V	3–12 V
Output load current	1–1000 mA	2–100 mA	1–100 mA	1–10 mA
Output capacitive loading	0.0005–10 μF	0.0005–0.02 μF	—	—
Output impedance, open-loop	30–5000 ohms	10–5000 ohms	100–5000 ohms	100–5000 ohms

3-11.2   Inverting Amplifier

An inverting ac amplifier is shown in Figure 3-24. From (3-34) the gain of the amplifier is approximately $-R_f/R_i$. To ensure a 6-dB/octave closure rate as discussed in Section 3-10.4 and to minimize high-frequency noise, a capacitor has been included in the feedback loop. The upper cutoff (-3 dB) frequency is determined by

$$f_H = 1(/2\pi R_f C_f)$$

An input capacitor has been included such that the amplifier will not respond to very low frequencies. Since the input is operating at a virtual ground, the input resistance of the amplifier is R_i, and the resistor and capacitor combination form a RC high-pass filter with a lower cutoff (-3 dB) frequency of

$$f_L = 1/(2\pi R_i C_i)$$

It should be clear that if it is desirable that the amplifier respond to a dc voltage the input capacitor is eliminated. Also, for maximum high-frequency response the feedback capacitor could be eliminated as long as the stability criterion has been met.

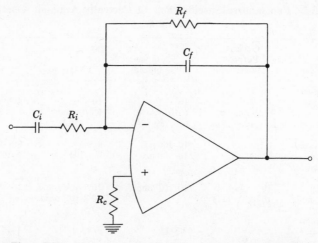

Figure 3-24 An operational amplifier used as an ac amplifier.

The output voltage of the amplifier chosen must be able to yield the desired voltage at the highest frequency of interest. Thus both output voltage and slewing rate are important. Additionally, the required output current, i_{out}, is

$$i_{out} = i_L + \frac{V_{out}}{Z_f}$$

since the circuit must power both the load and the feedback network Z_f. Input offset voltage and drift is specified at the amplifier input, and affects the output voltage by

$$\Delta e_o' = e_{os} \frac{R_f}{R_i}$$

where $\Delta e_o'$ is the output voltage due to offset voltage e_{os} and $-R_f/R_i$ is the closed-loop gain. Recall from Section 3-10.2 that the resistor R_c has been included to minimize errors caused by input bias currents. Any remaining offset current will affect the output by

$$\Delta e_o'' = i_{os} R_f$$

where i_{os} is the input offset current. Since the input of the amplifier is at virtual ground, the input voltage limit of the operational amplifier is not important. The total output voltage due to offset is, therefore,

$$\Delta e_o = \Delta e_o' + \Delta e_o''$$

3-11.3 Noninverting Amplifier

The noninverting amplifier is very useful for impedance matching. Recall that under certain conditions the input impedance is given by (3-44). For convenience (3-44) is repeated for resistive components. Thus

$$Z_{\text{in}} = Z_{io}\left(\frac{A(\omega)}{1 + R_f/R_i}\right) \qquad (3\text{-}49)$$

For maximum input impedance $R_f = 0$ (straight wire connection) and (3-49) gives

$$Z_{\text{in}} \approx Z_{io}A(\omega) \qquad (3\text{-}50)$$

From (3-42) we see that the closed-loop gain of the amplifier is unity. If amplification (gain) is desired we see from (3-42) and (3-49) that a trade off between input impedance and gain is necessary, all else being equal.

The input bias current and offset affects the output voltage $\Delta e_o''$ by

$$\Delta e_o'' = i_{os}R_i$$

However, offset voltage will be minimized by the inclusion of R_c. The input offset voltage affects the output voltage by

$$\Delta e_o' = e_{os}\frac{R_i}{R_c}$$

The total output voltage due to offset is, therefore,

$$\Delta e_o = \Delta e_o' + \Delta e_o''$$

3-11.4 Summing Amplifier

The summation of a number of input signals can be performed using the circuit shown in Figure 3-25. The choice of the values of the resistors $R_j(j = 1, 2, \ldots N)$ depends on the output impedance of the signal sources. Nominally, R_j would be much greater than the output impedance of the

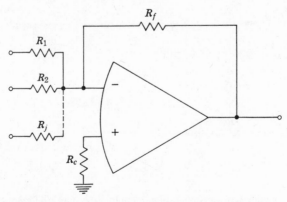

Figure 3-25 An operational amplifier used as a summer.

signal sources [recall (3-13)]. The output voltage from this circuit is

$$e_o = -R_f \sum_{j=1}^{N} \frac{e_j}{R_j} \tag{3-51a}$$

If the input resistors are all equal ($R_j = R$), then (3-51a) becomes

$$e_o = -\frac{R_f}{R} \sum_{j=1}^{N} e_j \tag{3-51b}$$

For the case $R_j = R$, the equivalent input resistance is R/N. Thus from (3-30) we see that

$$R_c = \frac{RR_f}{R + NR_f} \tag{3-52}$$

wherein we have assumed $R_o \ll R_f$. Equation 3-52 gives us the minimum offset voltage. The dc offset voltage at the output, due to the offset voltage of the amplifier, is

$$\Delta e_o' = \frac{N e_{os} R_f}{R}$$

The input bias current affects the output as

$$\Delta e_o'' = i_{os} R_f$$

Then, as before, the total voltage due to offset is

$$\Delta e_o = \Delta e_o' + \Delta e_o''$$

3-11.5 Differential Amplifier

The inverting and noninverting configurations may be combined to form a differential amplifier as shown in Figure 3-26. The gain of the amplifier is

$$e_o = \frac{R_2}{R_1} e_2 - \frac{R_f}{R_i} e_1 \tag{3-53}$$

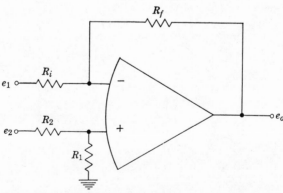

Figure 3-26 An operational amplifier used as a differential amplifier.

The maximum allowable input voltage, termed the common-mode voltage, is given by

$$e_{in_{max}} = \frac{e_2 R_1/R_2}{1 + R_2/R_1}$$

The input voltage e_1 does not matter since the other side of R_1 is at virtual ground [no current (ideally) flows into the amplifier at this point]. If $R_1 = R_i$ and $R_f = R_2$ then the offset current i_{os} affects the output voltage by

$$\Delta e_o' = i_{os} R_2 \frac{R_f}{R_i}$$

The input offset voltage affects the output voltage by

$$\Delta e_o'' = e_{os} \frac{R_f}{R_i}$$

The total offset output voltage, therefore, is

$$\Delta e_o = \Delta e_o' + \Delta e_o''$$

One additional specification is required for a differential amplifier that has not been used previously: common-mode rejection (CMR). If $e_2 = e_1$, then (3-53) states that the output voltage should be zero. This will never be the case, and the common-mode rejection ratio expressed in decibels, is

$$CMR = 20 \log_{10} \frac{e_o}{e_1} \text{ dB} \qquad (3\text{-}54)$$

By definition, (3-54) applies for the case $R_1 = R_i$, $R_f = R_2$, and $e_1 = e_2$.

3-11.6 Comparator (Schmitt Trigger)

It is often useful to be able to determine if a voltage is greater than some reference voltage. A circuit that performs this operation is termed a comparator or Schmitt trigger and is shown in Figure 3-27. The circuit obeys the

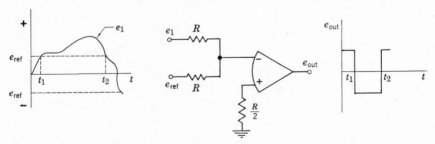

Figure 3-27 An operational amplifier used in a comparator configuration.

following relations

$$e_{\text{out}} = +\text{supply voltage}; \quad e_1 + e_{\text{ref}} < 0$$
$$e_{\text{out}} = -\text{supply voltage}; \quad e_1 + e_{\text{ref}} > 0$$

Although the operational amplifier configuration functions as a comparator, its switching time is determined by the slewing rate of the amplifier. This switching time can never be as fast as a circuit constructed as a triggering circuit only. The operation of the circuit is obvious by considering the fact that there is no feedback loop. The input impedance of the circuit with respect to ground is, from (3-37), R. As in previous circuits, R_c is connected to ground to minimize the offset voltage.

The error in the triggering level, referred to the input, is

$$e_{\text{error}} = Ri_{os} + 2e_{os}$$

The factor of 2 is present because of the voltage divider between e_1 and e_{ref}.

3-11.7 Integrator

Consider the integrator configuration shown in Figure 3-28, with C_i and R_f removed. The current through R_i is e_{in}/R_i. Since the minus input is at virtual ground, the current at the output is

$$\frac{e_{\text{in}}}{R_i} = -C_f \frac{de_{\text{out}}}{dt}$$

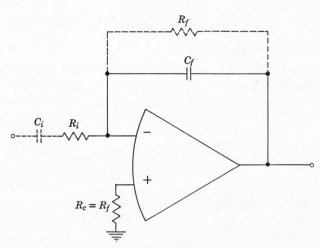

Figure 3-28 An operational amplifier used in an integrator configuration.

where e_{out} is the output voltage. Hence integrating this yields

$$e_{out} = -\frac{1}{R_i C_f} \int_0^T e_{in}\, dt$$

If e_{in} is a constant, we find that $e_{out} = -e_{in}T/R_iC_f$. If $e_{in} = A_0 e^{i\omega t}$, then $e_{out} = -1/(j\omega R_i C_f)$. Notice that this last expression for e_{out} can also be obtained from (3-34). Furthermore, an operational amplifier can be used to integrate a constant voltage (dc) whereas the low-pass filter, as described in Section 2-3.2, cannot.

The output voltage due to offset voltage, e_{os}, and offset current, i_{os}, is

$$\Delta e_o = \frac{e_{os}R_f'}{R} + i_{os}R_f'$$

where R_f' represents the equivalent leakage resistance of the capacitor. For the greatest leakage resistance polystryene capacitors should be used. The input offset voltage appears directly at the output; that is, it is not multiplied by the closed-loop gain of the amplifier. However, any dc drift will be integrated and cause a slow drift toward saturation. A feedback resistor, R_f, is thus included in the integration configuration to avoid saturation. An input capacitor can also be included, if desired, to avoid integration of any dc voltage that might be applied to the input. Because of the input and feedback capacitor the frequency range which the amplifier performs as an integrator is

$$\frac{1}{2\pi R_i C_i} \ll f \ll \frac{1}{2\pi R_f C_f}$$

Notice that in this case $R_c = R_f$ for minimum voltage offset.

3-11.8 Differentiator

If the input impedance of an operational amplifier is that for a capacitor C and the feedback impedance is that for a resistor R, then (3-34) may be written as

$$\frac{e_0}{e_{in}} = j\omega RC$$

For a harmonic input signal this is recognized as differentiation. Note that unlike the RC filter, differentiation can be accomplished with an increase in amplitude. However, maximum gain is at high frequencies, thus high-frequency noise is amplified more than low-frequency noise. Of a more serious nature is the fact that the rate of closure is approximately 12 dB/octave, making the differentiator inherently unstable in operation. One method of reducing noise and preventing instability is the addition of an input resistor. At high frequencies the capacitive impedance is negligible and the circuit operates as an amplifier with resistive feedback. A capacitor in the feedback loop would also accomplish the same purpose and is generally included for further reducing high-frequency noise. The addition of the input and feedback capacitors makes the differentiating circuit exactly the same as the integration circuit, only the frequency range is changed. That is, the frequency range is

$$0 \leqq f \ll 1/(2\pi RC)$$

where $R_i C_i = R_f C_f = RC$. The maximum frequency at which the circuit may be used depends upon the slewing rate of the amplifier.

As a final remark we note that any noise impressed upon the signal being differentiated will be accentuated (as opposed to smoothed, by integration) and consequently appear as amplified noise in the output voltage. For this reason differentiation circuits are, in general, seldom used for analog computers for solving differential equations, although it is feasible theoretically. They can be used, with discretion, for simple differentiation. For example, if a velocity transducer were used and a voltage proportional to acceleration was desired, an operational amplifier could be used. The advantage of using an operational amplifier instead of an RC filter is that an operational amplifier permits voltage gain and impedance matching.

3-11.9   Charge Amplifiers

As was discussed in Application 3-2, capacitive transducers are often employed to measure displacement and acceleration. Consider the circuit shown in Figure 3-29 wherein a dc voltage is placed in series with a capacitance $C_1 + \Delta C_1$. A change in capacitance causes a change in charge across the feedback capacitor C, which in turn results in a change in the output

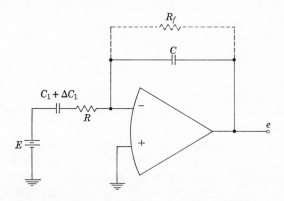

Figure 3-29   Equivalent circuit of a charge amplifier.

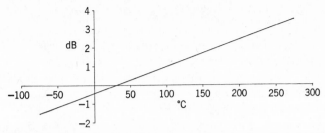

Figure 3-30 Typical variation in capacitance of a piezoelectric accelerometer as a function of temperature.

voltage e. Thus

$$q = C_1E = -Ce$$

and the increment of change of the output voltage is

$$-de = \frac{E}{C} dC_1 + \frac{C_1}{C} dE$$

Note that either, or both, a change in C_1 or a change in E, will cause a change in the output voltage. In either case a change in charge has taken place, and hence the amplifier is termed a charge amplifier. Piezoelectric transducers do not have an impressed voltage but do produce a change in charge, $dq = -Cde$.

The circuit has the property of being independent of shunt capacitance across the input. The independence of input capacitance permits the use of long shielded cables between the transducer and the amplifier without significantly affecting the accuracy of the conversion of charge to voltage. However the total capacitance, including the equivalent capacitance of the transducer and cable, must remain constant. Figure 3-30 shows the variation in capacitance as a function of temperature for a typical piezoelectric accelerometer. Thus if the temperature of the transducer is changed, the charge amplifier must be used with caution. Since the output voltage of the amplifier is inversely proportional to the value of the feedback capacitance, the smallest value that will be large compared to stray capacitance should be used for maximum output voltage. The inclusion of R_f is necessary so that the amplifier offset current will be sufficiently small to prevent saturation. Recall the discussion of the low-frequency limitations of the integration configuration. The lower limiting (cutoff) frequency of the charge amplifier is

$$f_L = 1/(2\pi R_f C)$$

3-11.10 Automatic Gain-Control (AGC) Amplifier

It is often useful to provide a constant voltage output from an amplifier or to be able to control the gain of an amplifier with an external voltage. It is desirable for the gain to be controlled regardless of any change in input voltage. An amplifier that has this capability is called an automatic gain-controlled amplifier or a compressor amplifier. This amplifier is discussed again in Section 6-2.2 and Application 6-1.

In order to obtain automatic gain control the basic inverting amplifier configuration is modified by the use of a field effects transistor (FET). The equivalent resistance between the drain and source of a FET can be approximated by

$$R_{\mathrm{DS}} = \frac{R_0}{1 - V_{\mathrm{GS}}/V_p} \tag{3-55}$$

where V_{GS} is the gate-to-source bias voltage, V_p is the gate-to-source bias for zero conduction, and R_0 is the value of the channel resistance at $V_{\mathrm{GS}} = 0$. These characteristics are found from the data sheet supplied by the manufacturer. Note that for small ac signals applied between the two ports of the FET (drain and source), the device presents a resistance with a value determined by the gate-to-source voltage. In order to minimize distortion a low equivalent resistance of the FET must be used.

The FET is used in the inverting configuration as shown in Figure 3-31. The gain of the amplifier is $-R_f/R_i$, where R_i in this case is given by (3-55). R_i is the drain-source resistance of the FET and is controlled by V_{GS}, the gate-to-source voltage. This is possible since the FET source is always at ground due to the fact that the negative input of the amplifier is at virtual ground. Typical variations in amplifier gain with gate voltage is shown in Figure 3-32.

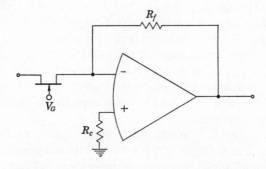

Figure 3-31 Simplified automatic gain-control amplifier.

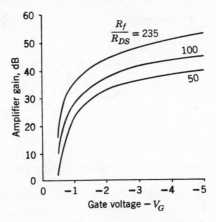

Figure 3-32 Amplifier gain versus gate voltage for a compressor amplifier using a FET.

The device is somewhat temperature sensitive, that is, over a temperature range of 50°C the change in gain is typically ±1 dB. At higher input signal levels, the distortion condition described earlier begins to become a problem. The distortion problem can be overcome by adding a resistor in shunt across the FET. Then when the control range of the FET is exceeded, the amplifier performs normally. If the amplifier is used to provide an output voltage that is a constant regardless of the input, as opposed to the gain controlled from an external program voltage, the output voltage is rectified and integrated (see Chapter 4) and applied to the gate of the FET. The range of the constant

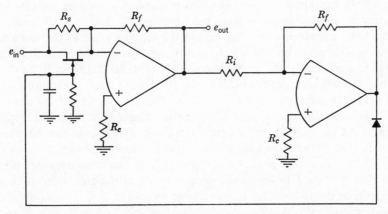

Figure 3-33 Functional diagram of a compressor amplifier using a FET and operational amplifiers.

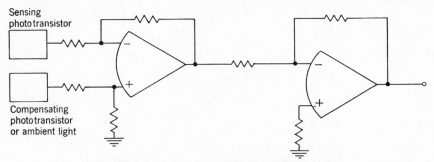

Figure 3-34 Functional diagram of a nephelometer.

gain can be improved considerably by the addition of a second amplifier for amplifying the control signal. In this case a 2-dB change in output level can be maintained for a 30-dB change in the input. The complete circuit with shunt resistor and second stage of amplification is shown in Figure 3-33.

Application 3-5   A Deep Ocean Nephelometer

A nephelometer is a device for measuring the amount of suspended material in a fluid. Suspended material is used by the oceanographer to trace flow patterns and by ecologists to measure pollution.

The block diagram of a nephelometer is shown in Figure 3-34. A collimated beam of light is projected into the water through a window. For constant intensity of the light source a constant-current constant-voltage power supply (as will be discussed in Section 3-17.2) is used to power the light. In general, light transducers are temperature sensitive and for this reason two have been used with a differential amplifier. The light transducers (phototransistors) are mounted in a disk of aluminum together with the light collimator. Phototransistors were used because of their lack of hysteresis. Since both transducers are mounted in the disk they are at the same temperature. One phototransistor measures the amount of light reflected back by the fluid medium and the other phototransistor is shielded from all light. The second transistor output voltage is only a function of the temperature and not the light reflected back from the medium; thus the output voltage from the differential amplifier is proportional to reflected light only. A second amplifier is included to increase the magnitude of the signal from the differential amplifier. In this case the instrument was designed for the deep ocean where ambient light is not a problem. However, if the compensating phototransistor was used to measure ambient light and the collimated light was greater than the ambient light, such a system could be used for shallow water. The output signal is multiplexed to the surface ship by a technique described in Section 6-2.3.

Application 3-6   *Differential Amplifier Used for Strain Gauges*

One of the simplest transduction mechanisms is employed in the variable-resistance transducer. The electrical resistance of any conductor is a function of its dimensions and temperature. When the dimensions or temperature are varied mechanically as a fixed, steady current is passed through the conductor, the voltage developed across the conductor varies as a function of this mechanical excitation. Alternatively, a fixed voltage can be applied across the conductor, and the current which flows varies as a function of the applied mechanical excitation. These relations follow directly from Ohm's law.

If the temperature is held constant, the change in resistance can be used to measure the strain in a mechanical element. The change in resistance ΔR is

$$\frac{\Delta R}{R} = \text{GF}\frac{\Delta L}{L} \tag{3-56}$$

where L is the length and R the resistance of the unstrained conductor, ΔL its change of length, and ΔR its change in resistance caused by ΔL. The dimensionless magnitude GF is termed the gauge factor and varies for different metals between -12.0 and $+4.0$. On the other hand semi-conductor strain gauges have gauge factors which vary between ± 120.

Consider a circuit, termed a bridge circuit and shown in Figure 3-35, which can be designed to give an output voltage only when the resistances change. It can be shown that the output voltage e_{ac} is

$$e_{ac} = e\left(\frac{R_1}{R_1 + R_4} - \frac{R_2}{R_2 + R_3}\right)$$

Note that when $R_1/R_4 = R_2/R_3$, $e_{ac} = 0$ and the bridge is said to be balanced. Let R_1 change by a small amount ΔR_1. Then the output voltage is

$$e_{ac} = e\left(\frac{R_1 + \Delta R_1}{R_1 + R_4 + \Delta R_1} - \frac{R_2}{R_2 + R_3}\right)$$

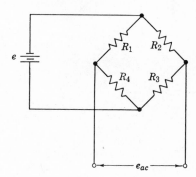

Figure 3-35   A four-arm bridge circuit.

It is seen that, in general, e_{ac} is not directly proportional to the change ΔR_1. However, if $R_4 \gg R_1$, the output voltage may be written as

$$e_{ac} \approx e\left(\frac{\Delta R_1}{R_1 + R_4} - \frac{R_2}{R_2 + R_3} + \frac{R_1}{R_1 + R_4}\right) \tag{3-57}$$

Hence the change in e_{ac} is now proportional to ΔR_1 since the second and third terms in (3-57) are constant for a given circuit. Note, however, that the amplitude of e_{ac} will be extremely small since $R_1/R_4 \ll 1$. Therefore, either the bridge is again balanced and the amount of change in R_3 is measured or an amplifier is used. In the former method the amount of change in R_3 is proportional to the strain; in the latter method the change in voltage is. Another way to obtain an output voltage that is directly proportional to R_1 is to let $R_1 = R_2 = R_3 = R_4 = R$ and $\Delta R_1 = -\Delta R_2 = \Delta R_3 = -\Delta R_4 = \Delta R$. Then (3-57) becomes

$$e_{ac} = e\frac{\Delta R}{R} \tag{3-58}$$

In some cases it may be necessary to use a grounded power supply to power the bridge. Since the input of the amplifier is also grounded a ground loop results. Ground loops are discussed in Appendix C. If the power supply is not grounded noise may be induced due to lead length. The two cases of noise pickup with a differential amplifier are shown in Figure 3-36. The signal leads and the internal capacitances are shown lumped as C_d. A guard shield providing an electrostatic shield around the input circuitry breaks the stray capacitance into two series capacitances, C_d and C_g. A much higher impedance is then presented to the flow of common-mode signals and thus the noise is subtracted since the signal plus noise is connected to the inverting input and only the noise is connected to the noninverting input.

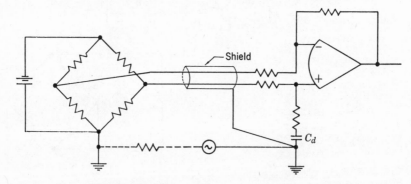

Figure 3-36 A differential amplifier used in a common-mode reject configuration to reduce noise.

3-12 Operational Amplifiers Used with Nonlinear Elements

3-12.1 Introduction

It is often useful to have an amplifier with nonlinear rather than linear response. For example, when computing the mean-square value of a signal, the input signal must be squared and then integrated. Thus a squaring operation must be made. Nonlinear elements can be classified as log amplifiers or diode function generators. The log amplifiers provide an output voltage proportional to the logarithm of the instantaneous value of the input voltage. Diode function generators generally consist of a number of diodes and the nonlinearities are made adjustable by some means, such that a piecewise linear approximation to the desired nonlinear gain function is made.

3-12.2 Logarithmic Amplifiers

Analog logarithmic amplifiers can be designed using several different circuit techniques, but the most common designs use the logarithmic characteristics of a semiconductor junction. If the base and collector of a forward-biased transistor are held at 0 volts, the base-emitter voltage will be proportional to the logarithm of the collector current. It can be shown that the transfer conductance of an NPN silicon transistor is

$$I_C = -\alpha_n I_{ES} e^{qV_e/kT}$$

where I_C is the collector current, α_n is the forward short-circuit current gain, I_{ES} is the emitter reverse saturation current, q is the charge of one electron, V_e is the emitter-base voltage, k is Boltzmann's constant, and T is the absolute temperature. If the transistor is used as the feedback element in an operational amplifier and R_i is used as the input, then the transfer function is

$$E_{\text{out}} \approx -\frac{kT}{q} \log_e \frac{E_i}{R_i} + \frac{kT}{q} \log_e (I_{OS}) \qquad (3\text{-}59)$$

where I_{OS} is the offset current. The circuit is shown in Figure 3-37.

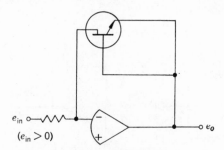

Figure 3-37 An operational amplifier used as a logarithmic amplifier.

As can be seen from (3-59) the circuit does provide a logarithmic response, but the response is very temperature dependent and there is an offset term due to the offset current I_{OS}. Therefore, additional circuit components are normally added to stabilize the gain and offset variation. Also, components are added to allow the subtraction of the offset term and for scaling purposes. The exact method of providing temperature-drift compensation and offset null are unimportant for purposes here. It is sufficient to note that many manufacturers have log devices for use with operational amplifiers. These devices typically have a dynamic range of 40 to 60 dB and obey the relationship

$$E_{\text{out}} = K_1 \log_e (K_2 E_i)$$

where E_i is the input voltage and K_1 and K_2 are scaling constants which, in general, may be set externally.

3-12.3 Diode Function Generators

Diode function generators (DFG) use diodes to form some desired nonlinear gain function. In general, a number of diodes are used such that the DFG can provide a piecewise linear approximation to the desired gain function. When used with adjustable linear approximations, the DFG is sometimes called an arbitrary function generator. The basic form of a DFG is shown in Figure 3-38 together with approximate equations for the slopes and

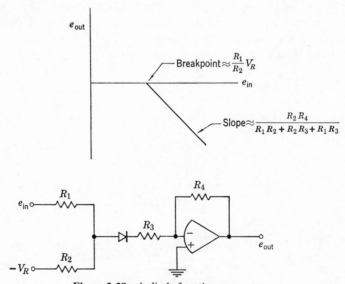

Figure 3-38 A diode function generator.

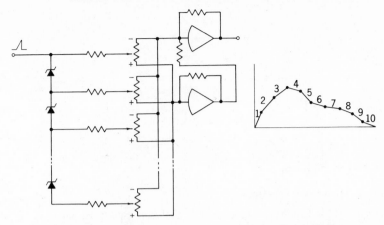

Figure 3-39 A diode function generator using Zener diodes.

breakpoints. The characteristic shown assumes an ideal diode; actually, there will be some rounding of the breakpoint and some small offset in the breakpoint. The basic concept can be expanded considerably to form more breakpoints with corresponding different slopes. The output will then be the composite sum of input currents from each diode-resistor combination. It should be noted that unless an operational amplifier is used with each diode there will be some interaction between the diodes due to an impedance mismatch; for example, the first break point will affect the second breakpoint.

Cascading of diode networks is illustrated in a modified circuit shown in Figure 3-39. This circuit uses Zener diodes to form the breakpoints. Zener diodes will not conduct below some voltage; however, when the voltage is increased to some value, termed the Zener voltage, the diode will conduct. Since two amplifiers are used in the circuit, negative as well as positive slopes can be generated. Negative and positive slopes could be generated with the diode circuit previously described by the use of more than one amplifier at each breakpoint.

3-13   Applications Using Operational Amplifiers with Nonlinear Elements

3-13.1   Introduction

Operational amplifiers with nonlinear elements may be used for obvious applications such as logarithmic conversion and squaring but they may be used for other operations such as multiplication, antilogs, and square-root transformations. These operations are considered in this section.

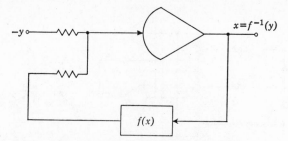

Figure 3-40 A circuit that generates the inverse of a function.

3-13.2 Inverse Operations

Given that a functional relationship between two variables is $y = f(x)$, then the inverse relationship is defined as $x = f^{-1}(y)$. A circuit which can generate a particular functional relationship can also generate the inverse by solving the implicit-function equation

$$f(x) - y = 0$$

Such a circuit is shown in Figure 3-40. Thus if a circuit can generate a given function $f(x)$, it can generate the inverse by placing the function generator in the feedback loop of an operational amplifier. Then, if a variable $-y$ is applied at the input of the amplifier, the output will be $f^{-1}(y)$.

For example, if it were desired to generate a variable raised to a function power, say $z = x^{a/b}$, where a and b are both integers and $b > a$, then we must generate $z = x^{1/b}$ or $z^b = x$. In so doing all the terms $x^{2/b}, x^{3/b}, \ldots$ $x^{(b-1)/b}$ are also obtained. The circuit is slightly different when b is an even or an odd integer, as can be seen in Figure 3-41. As a specific example the circuit in Figure 3-42 will generate $z = x^{1/3}$.

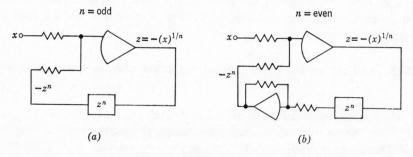

Figure 3-41 (a) Generation of fractional exponents when n is odd, and (b) when n is even.

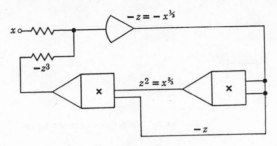

Figure 3-42   Generation of $x^{1/3}$ and $x^{2/3}$.

3-13.3   Multiplication and Division

In general, multiplication uses what is termed the quarter square technique as shown in Figure 3-43. In this case two squaring circuits are used, one to square the quantity $(x + y)$ and the other to square the quantity $(x - y)$. The outputs of the two squaring circuits are then scaled and summed such that

$$\tfrac{1}{4}[(x + y)^2 - (x - y)^2] = xy$$

In general, the input is scaled such that the output is $xy/10$. The scaling is necessary to use the full dynamic range of the two input amplifiers without saturation of the output amplifier.

By placing the squaring network in the feedback loop as shown in Figure 3-44, division can be obtained. In this case the input to the summing junction

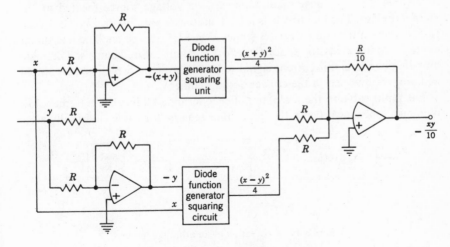

Figure 3-43   A quarter-square multiplier.

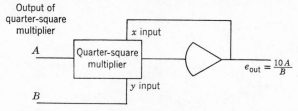

Output of quarter-square multiplier

Figure 3-44 Division using a quarter-square multiplier.

is $xy/10 + z = 0$; therefore, the output is $x = -10z/y$, which is valid for $y \neq 0$.

3-13.4 Generation of an Arbitrary Power of the Input

By the combination of a log amplifier, multiplier, and inverse log circuit, arbitrary generation of a power of x may be obtained. In the block diagram of the circuit shown in Figure 3-45, the operational amplifiers associated with the nonlinear function generators have been omitted for simplicity.

3-14 Zero Drive Amplifier*

A Zero Drive amplifier is an impedance matching amplifier for capacitance signal generators. Referring to Figure 3-46 it is seen that the signal originating at the transducer is fed into an electronic converter located in the transducer or in an external "line driver" placed close to the transducer. The transducer signal is converted to an electrical current which is derived, through a two-wire cable, from a substantially constant dc voltage source located at the main amplifier. The line driver is a FET used as a variable resistance. Recall (3-55) wherein the equivalent resistance of a FET was given. The line driver contains components to reduce the effect of temperature changes. The main amplifier converts this current signal to a voltage signal which is then amplified and conditioned in the conventional manner.

The input cable to the main amplifier is terminated into an extremely low dynamic impedance; hence the impedance seen by the cable is virtually zero.

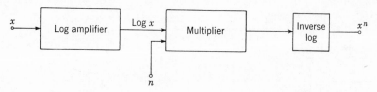

Figure 3-45 Generation of an arbitrary power.

* The Zero Drive is a patented device of MB Electronics, New Haven, Conn.

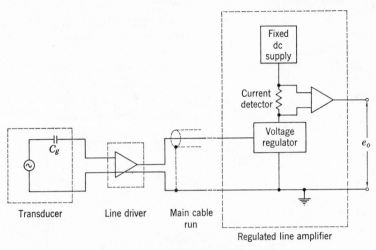

Figure 3-46 Typical zero drive system.

This results in a system that has a very low sensitivity to electrical noise pickup and triboelectric noise generated in the cable due to cable motion. The low noise level allows signals to be measured over a large range with an excellent signal-to-noise ratio that is almost independent of cable capacitance.

The system is referred to as a Zero Drive system because the cable is effectively driven with zero signal voltage. Hence no current is required to drive the cable capacitance. As a result, cable capacitance has a negligible effect on the maximum output voltage and frequency response.

3-15 Active Filters

An active filter is a collection of judiciously arranged passive elements (resistors, capacitors) and at least one active element, usually an operational amplifier. Although not specifically mentioned in Chapter 2, the passive filter has several disadvantages; namely, low-input and high-output impedances, insertion loss, and relatively large physical size at low frequencies (< 200 Hz). An active filter overcomes these drawbacks. Because it uses an operational amplifier, impedance matching is usually no problem because of the intrinsic high-input–low-input impedance of the operational amplifier configuration. Insertion loss is eliminated because the amplifier configuration can produce gain, if desired. Also, since the active filter uses just resistors and capacitors its overall physical size can be greatly reduced, even at very low frequencies. The active filter, however, does introduce other errors and limitations. First the dynamic range of the filter will be limited to that of the operational amplifier at the particular frequencies of the filter which, in general, is less than that of an equivalent passive filter. Furthermore, the active filter introduces noise to the signal because of the active elements in the operational amplifier.

In the actual design of active filters there are four different classifications of circuits. Regardless of which type of active filter selected, the resulting network, consisting of RC components and an operational amplifier, simulates effectively one of the following three transfer functions:

1. Low-pass filter:

$$|H(\omega)|^2 = \frac{A_0^2}{D_0}$$

2. Bandpass filter:

$$|H(\omega)|^2 = \frac{A_0^2(\omega/\omega_0)}{D_0}$$

3. High-pass filter:

$$|H(\omega)|^2 = \frac{A_0^2(\omega^2/\omega_0^2)}{D_0}$$

where

$$D_0 = \left[\left(1 - \frac{\omega^2}{\omega_0^2}\right)^2 + \frac{\alpha^2\omega^2}{\omega_0^2}\right]$$

and A_0 is the gain.

The quantity ω_0 in the above transfer functions is the cutoff frequency of the low- and high-pass filters and the center frequency of the bandpass filter. The quantity α is a constant, chosen to give certain desired properties of $H(\omega)$ in the passband. For example, in the case of the low-pass filter, if $\alpha = \sqrt{2}$, $|H(\omega)|^2$ reduces to (2-73) with $n = 2$ (i.e., 12-dB/octave selectivity) and we have the case of a maximally flat (Butterworth) response· A similar choice of α for the high-pass filter also yields a maximally flat response. On the other hand, for $\alpha > \sqrt{2}$ the selectivity is decreased and, conversely, for $\alpha < \sqrt{2}$ it is increased. In the case of a bandpass filter $\alpha = 1/Q(Q > 10)$ wherein Q is given by (2-40). When the filters are cascaded for more selectivity,

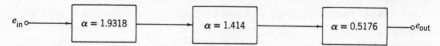

Figure 3-47 Cascaded active filters to obtain a sixth-order Butterworth filter.

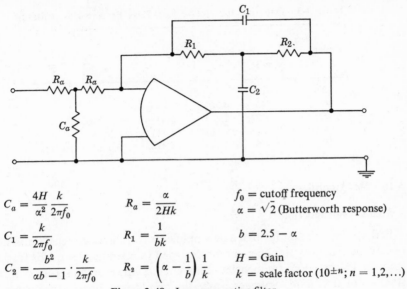

$$C_a = \frac{4H}{\alpha^2}\frac{k}{2\pi f_0} \qquad R_a = \frac{\alpha}{2Hk}$$

$$C_1 = \frac{k}{2\pi f_0} \qquad R_1 = \frac{1}{bk}$$

$$C_2 = \frac{b^2}{\alpha b - 1}\cdot\frac{k}{2\pi f_0} \qquad R_2 = \left(\alpha - \frac{1}{b}\right)\frac{1}{k}$$

f_0 = cutoff frequency
$\alpha = \sqrt{2}$ (Butterworth response)
$b = 2.5 - \alpha$
H = Gain
k = scale factor ($10^{\pm n}$; $n = 1,2,\ldots$)

Figure 3-48 Low-pass active filter.

each stage has a unique α. These values are given in Table 3-3. A typical cascade for a sixth-order Butterworth low-pass filter is shown in Figure 3-47. To further illustrate how an active filter is designed and how the values of ω_0 and α correspond to the RC components, a low-pass and high-pass active filter are shown in Figures 3-48 and 3-49, respectively. Other types of active filter configurations are given in refs. 13 and 18.

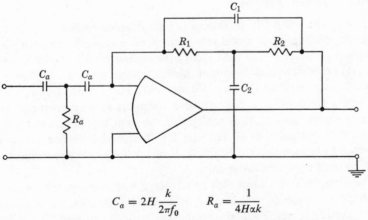

$$C_a = 2H\frac{k}{2\pi f_0} \qquad R_a = \frac{1}{4H\alpha k}$$

Figure 3-49 High-pass active filter (see Figure 3-48 for definition of symbols).

Table 3-3   Damping for Active Low-Pass Butterworth Filters

Order n	α_j				
	$j = 1$	2	3	4	5
2	1.414	—	—	—	—
4	1.845	0.7654	—	—	—
6	1.932	1.414	0.5176	—	—
8	1.962	1.663	1.111	0.3896	—
10	1.976	1.783	1.414	0.9081	0.3128

3-16   Dc Amplifiers

3-16.1   Introduction

Perhaps the simplest form of a dc amplifier is two-direct-coupled transistors in cascade. Such an amplifier can be designed to operate over the bandwidth from dc to very high frequencies. This circuit, however, is sensitive to the stability of the supply voltages, since the amplifier is unable to distinguish between a 1-mV signal and a 1-mV change in supply potential. Changes in tube characteristics with voltage or time, or transistor parameters with temperature, are other causes of instability and drift in direct-coupled amplifiers. Occasionally attempts are made to provide external balance controls to compensate for the above problems. Such a circuit is successful in many vacuum-tube voltmeters where frequent manual adjustments can be made to balance out the drift; however, employment of such a device in sensitive amplifiers requiring long-term stability is not satisfactory.

3-16.2   Chopper Amplifier

One solution to the drift problem in direct-coupled amplifiers is to "chop" the slowly varying input signal into a form of ac, amplify this with a conventional ac amplifier, and then rectify the output back to a slowly varying (dc form) proportional to the input. Such a system is shown in Figure 3-50. In the chopper amplifier shown in Figure 3-50 a mechanical chopper employs a polarized (responsive to only positive voltages) ac magnet to operate the switch arm and contacts, thus performing the switching in synchronism with the ac supply (usually 50–60 Hz, the line frequency). Hence the input is connected to the amplifier only one-half of the time. An electronic chopper could be used for much faster rates.

In Section 7-3 it is shown that any voltage with frequencies greater than the chopper frequency will appear as lower frequencies. Such errors are termed alaising errors. To minimize alaising errors a low-pass filter is formed by $R_1 C_1$ at the input. To ensure sufficient attenuation of the high-frequency

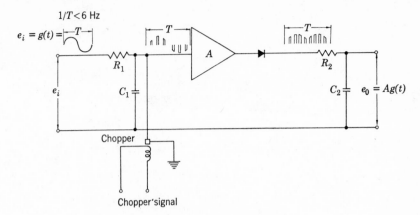

Figure 3-50 Chopper amplifier.

components the cutoff of the low-pass filter is chosen to be one tenth of the chopper frequency. Thus the bandwidth of the amplifier is limited to 6 Hz if a 60-Hz line current is used to drive the chopper.

As indicated in Figure 3-50, the peak-to-peak amplitude of the amplifier input will equal the input voltage during the half cycle in which the chopper is open, and therefore, any change in the input voltage during this half period will only affect the output during this period. The output of the amplifier is rectified by diode D, and harmonics of the chopping frequency are removed by the low-pass filter formed by R_2C_2 at the output.

3-16.3   Chopper-Stabilized Amplifier

Higher frequency response can be obtained with the chopper-stabilized amplifier when compared to the chopper amplifier. The block diagram of the chopper-stabilized amplifier is shown in Figure 3-51. The chopper-stabilized amplifier uses a chopper amplifier of gain A_2 and a dc amplifier of gain A_1. Resistors R_1 and R_2 and capacitor C_2 form a low-pass filter for the reasons discussed in the previous section. The resistor R_3 and the capacitor C_1 form a high-pass filter so that only high frequencies are directly connected to the dc amplifier. The diode and low-pass filter on the output of the chopper amplifier are used for the same reasons discussed previously.

The voltage on C_2 is chopped, amplified in the ac amplifier, rectified, filtered, and applied as additional input to the dc amplifier as a drift correcting voltage. Again, although a mechanical chopper is shown, an electronic chopper could be used as well. This input cancels the equivalent input voltage of the dc amplifier and brings the drift output voltage back to nearly zero. At the same time, the higher frequency components are passed through C_1

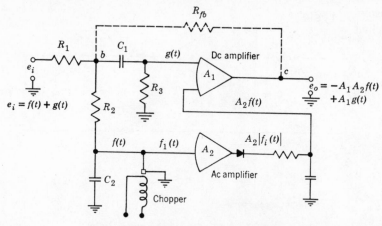

Figure 3-51 Chopper-stabilized dc amplifier.

directly into the dc amplifier. The positive input of the operational amplifier is used for the input of the chopped signal. This loss of the positive input confines chopper-stabilized amplifiers to circuits based on the inverting configuration.

If e_d is the drift voltage of the dc amplifier, then the output due to drift is

$$e_{od} = -A_1 e_d$$

where A_1 is the gain of the dc amplifier. Note that the output is 180° out of phase with the input. With the circuit of Figure 3-51 and an ac amplifier gain of A_2, the output for low frequencies will be

$$e_{o_{low}} = A_1 A_2 e_i$$

while the output for all frequencies passing through C_1 and the dc amplifier only (high frequencies) will be

$$e_{o_{high}} = A_1 e_i$$

The overall frequency response is shown in Figure 3-52. If now an input resistor is added in series to the input at b of Figure 3-51 and a feedback

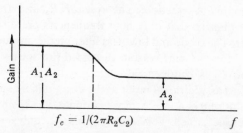

Figure 3-52 The open-loop gain of the chopper-stabilized dc amplifier shown in Figure 3-51.

resistor between points b and c are added, then the gain of the chopper amplifier depends only on the passive elements and the gain will be a constant. Also notice that all drift will occur in the last amplifier. Therefore (3-28b) gives

$$\Phi_d = - \frac{\Phi_{d2}}{A_1 A_2 \beta}$$

where Φ_{d2} is the drift in the second amplifier. Thus the drift has been reduced by a factor of $1/(A_1 A_2 \beta)$.

3-17   Power Supplies

3-17.1   Introduction

Electronic power supplies can be defined as circuits which transform electrical input power (either ac or dc) into output power (either ac or dc). Electronic power supplies may be subdivided into four classifications: ac to ac (line regulators and variable frequency supplies), dc to dc (converters), dc to ac (inverters), and ac to dc. The last classification is the one most commonly used for instrumentation. The discussion in this section will, therefore, only be concerned with the ac to dc power supplies.

The simplest type of ac to dc power supply is a simple rectifier, arranged in a half-wave or full-wave bridge. A low-pass filter follows the rectifier. The RC time constant of the filter is extremely long so that the build-up (or build-down) time is large compared to a period of the ac voltage. The output of the filter then is a slowly varying dc voltage. The residual ac component which is superimposed on the dc output is termed ripple.

Most applications need a regulated power supply which requires a control element between the filter and the load. The purpose of the control element is to provide a buffering effect so that the voltage output is independent of load variations and input voltage changes. It is the purpose of the regulator to minimize all changes; that is, changes in the output voltage due to changes in the input voltage, output current, and temperature. Various types of regulated power supplies are described in the following.

3-17.2   Constant-Voltage/Constant-Current Supply

An ideal constant-votage power supply would have a zero output impedance at all frequencies. Thus the voltage would remain constant in spite of any changes in output current demanded by the load. Frequency response of a power supply is measured in terms of the transient recovery time or recovery time of the power supply. That is, the time required for the output voltage to return to a given dc output level following a sudden change in the load.

The ideal constant-current power supply would exhibit an infinite output

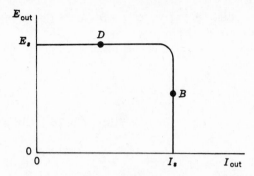

Figure 3-53 Operating curve of a constant current-constant voltage power supply.

impedance at all frequencies. Thus the ideal constant-current power supply would accommodate a load resistance change by altering its output voltage by the necessary amount to ensure that its output current would remain constant,

We see that the constant-voltage supply requires zero output impedance and a change in the output current whenever the load resistance changes, whereas the constant current supply requires an infinite output impedance and a change in the output voltage to any load resistance change. These two requirements cannot be met simultaneously. For any given value of load resistance, the power supply must act as either a constant-voltage source or a constant-current source. The constant-voltage-constant-current characteristics are shown in Figure 3-53. With no load attached $R_L = \infty$, $I_{out} = 0$, and the output voltage is equal to a predetermined voltage. When a resistive load is applied to the output terminals of the power supply, the output current increases, while the output voltage remains constant. Point D thus represents a typical constant-voltage operating point. Further decreases in load resistance result in increases in I_{out} with no changes in the output voltage until the output current reaches I_s. At this point the supply changes its mode of operation and becomes a constant current source. Point B represents a typical constant-current operating point. Still further decreases in the load resistance result in output voltage decreases with no change in output current, until finally, with a short circuit across the output load terminals, $I_{out} = I_s$ and $E_{out} = 0$. Notice that full protection against any overload condition (voltage or current) is inherent in the power supply since no load condition will cause an output which lies outside the operating curve shown in Figure 3-53.

The equivalent circuit of a typical constant-current/constant-voltage power supply is shown in Figure 3-54. The reference voltage is generated in a temperature-compensated circuit and is connected to a voltage reference amplifier for isolation between other parts of the circuit. The output of

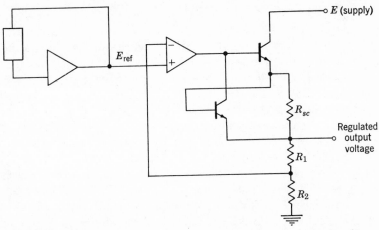

Figure 3-54 Typical equivalent circuit for a constant current-constant voltage power supply.

the reference amplifier is connected to the noninverting input of an error amplifier. The error amplifier is connected to the base of a transistor which increases the output impedance of the error amplifier. The emitter of the transistor is connected to the base of a second transistor and is called the current limiter The emitter is connected through a small current sense resistor to a voltage divider formed by R_1 and R_2. The divided voltage is connected to the inverting input of the error amplifier. The output of the power supply is thus

$$E_{out} = \frac{R_1 + R_2}{R_2} E_{ref}$$

Since the error amplifier is sensing the output voltage and comparing it to the reference voltage, the output voltage will remain at the predetermined value. The actual output leads of the power supply could be different than those to the error amplifier so that remote tracking of the output voltage could be accomplished. This is used for providing power to equipment away from the power supply.

The current limit transistor is connected to detect the voltage drop across the current sense resistor, R_{sc}. As the load current increases the voltage drop across the sense resistor turns "on" the current limit transistor and the transistor shunts current that would normally be available to the output circuit. The limiting load current is given approximately by $I_{limit} = 0.7/R_{sc}$ for silicon transistors. Either the constant-current section or constant-voltage section could be disconnected to provide either constant current or constant voltage.

3-17.3 Overvoltage Load Protection

If a series regulator fails, it usually becomes a short circuit rather than an open circuit. If failure occurs, the output voltage can then rise to the full rectifier value. Under this condition the normal current limit circuit is no longer operative, and the load current is limited by the load resistance.

A device to protect the load from overvoltages is termed a "crowbar." A circuit which operates independently of the power supply monitors the output voltage of the supply through a differential amplifier. If the output exceeds a preset voltage threshold a pulse is sent to a silicon-controlled rectifier (SCR) which is placed in parallel with the output of the supply. An SCR has the property that it does not conduct until the trigger pulses are sent to it. When a trigger pulse is sent to it, it is in the on state and offers a very low resistance Thus if the voltage to the comparator is greater than the preset level a pulse is sent to the SCR which then switches to its conducting state and shorts out the output of the power supply. Blocking circuitry is present so that the pulses are not turned off when the voltage falls to zero.

3-17.4 Dual Output Supply

Often it is required to use both a positive and negative dc power source having the same voltage and current capability (for example, with an operational amplifier). It might seem reasonable to meet such requirements using a single regulated dc power supply with a resistive voltage divider center-tapped to ground a shown in Figure 3-55. However such an arrangement results in an increase in the effective dc source impedance feeding each load. Assuming that the power supply has a zero output impedance, the impedance of the two legs are

$$\frac{1}{Z_1} = \frac{1}{R_1} + \frac{1}{R_2} + \frac{1}{R_{L2}}$$

$$\frac{1}{Z_2} = \frac{1}{R_1} + \frac{1}{R_2} + \frac{1}{R_{L1}}$$

where Z_1 and Z_2 are the impedances of the two legs, R_1 and R_2 are the two divider resistors, and R_{L1} and R_{L2} are the two loads. Thus a change in the

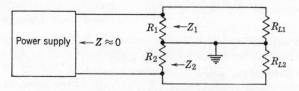

Figure 3-55 Center-tapped power-supply output.

current requirement of either load results not only in a change in its own dc voltage, but also in a change of the dc voltage feeding the other load. In nearly all cases, a simultaneous need for positive and negative dc voltages necessitates the use of two separate regulated power supplies.

One additional problem in powering operational amplifiers is that of driving a load at high frequency; that is, maintaining the desired voltages at the power-supply terminals of the operational amplifier for rapid changes in load. The line impedances become too great to be able to yield the instantaneous power. Therefore, it becomes necessary to supply the surge demands from a low impedance source such as a capacitor. It is important that the capacitor have a large Q, and should be either monolithic or Mylar type. The capacitor is placed physically as close to the amplifier power terminals as possible. The value of the capacitor can be determined from the current-voltage relations in a capacitor; that is,

$$I = C \frac{dV}{dt}$$

where I is the maximum instantaneous current into the load, dV is the maximum ripple allowed at the power supply terminal, and dt is the time interval of the transient.

3-17.5   Series and Parallel Operation of Power Supplies

In general, power supplies of the same manufacturer with the same model number can be operated in series or parallel. However, it is unusual for different model numbers to have this ability, principally due to the difficulties encountered with impedance matching.

3-18   Summary

The two primary functions of amplifiers are to amplify signals that are too low in level for their intended application and to isolate circuits from each other. Amplifiers can be broadly classified as either voltage amplifiers or power amplifiers. When used as a voltage amplifier the input impedance of the amplifier should be at least 100 times the output impedance of the device generating the input signal in order to introduce less than 1 % error in amplitude when both impedances are resistive. If the generator impedance is resistive and the input impedance of the amplifier is shunted with a capacitor, then in order for one to obtain broadest frequency response with minimum amplitude error, both the product of the shunt capacitance and the generator resistance and the ratio of the generator resistance to the amplifier input resistance must be much less than unity. If the generator impedance is capacitive, then the amplifier's shunt capacitance should be less than or equal to the generator capacitance and the input resistance should be very large so that the product of input resistance and the generator capacitance is much greater than unity to ensure a minimum amplitude error and lowest possible cutoff frequency. When using an amplifier as a control or power amplifier the output impedance of the amplifier and its load should be complex conjugates of each other for maximum power transfer.

Since an amplifier consists of active and passive elements, each of which is a noise generator, it is often useful to know the amount of noise the amplifier will contribute to the signal. The measure of this contribution is called the noise figure which is the ratio of the input signal-to-noise ratio to the output signal-to-noise ratio. When two or more amplifiers are used in cascade it is necessary, in order to keep the total noise figure to a minimum, to have the first amplifier with maximum gain and lowest noise figure and the other amplifiers at less than the maximum gain. When two or more amplifiers are cascaded the overall rise time is equal to the square root of the sum of the squares of the individual rise times of each amplifier. The overall bandwidth of the cascaded amplifiers, corresponding to the above increase in rise time, is decreased.

In addition to generating noise an amplifier will generate harmonic distortion when operated beyond its linear range. This nonlinear region of amplification is encountered when the input level to the amplifier passes a certain predetermined amount (usually obtained by specifying the amount of harmonic distortion one is willing to accept). The region between the noise level of the amplifier and the prescribed amount of harmonic distortion is called the dynamic range of the amplifier. Although the dynamic range of the amplifier will not change, the range of useful voltages it may amplify can be increased with the use of attenuators.

A very useful and common type of voltage amplifier is an operational amplifier. It is one which has very high-input impedance, low-output impedance, and large gain. It is almost invariably used with some amount of feedback which reduces distortion, drift, and fluctuation in frequency response. From its two basic configurations, inverting and noninverting, it can perform such linear and nonlinear operations as impedance conversion, summation, integration and differentiation, comparison, multiplication, division, and square roots. Furthermore, with various components in its feedback it can be formed into active filters, power supplies, and chopper-stabilized amplifiers.

References

1. W. Bennett, *Electrical Noise*, McGraw-Hill Book Company, Inc., New York, 1960.
2. P. Birman, *Power Supply Handbook*, Kepco, Inc., Flushing, New York, 1965.

3. T. Cate, "Designing with Nonlinear Function Modules," *EEE*, No. 6 (September 1969).

4. M. S. Ghausi, *Principles and Design of Linear Active Circuits*, McGraw-Hill Book Company, Inc., New York, 1965.

5. L. Levine, *Methods for Solving Engineering Problems Using Analog Computers*, McGraw-Hill Book Company, Inc., New York, 1964.

6. M. Millman and H. Taub, *Pulse, Digital, and Switching Waveforms* McGraw-Hill Book Company, Inc., New York, 1965.

7. C. Mullett, "Don't Buy Too Much Op Amp," *Eng. Design News (EDN)*, pp. 37–40, February 1, 1969.

8. G. E. Owens and P. W. Keaton, *Fundamentals of Electronics*, Vol. III Harper and Row, New York, 1967.

9. J. Rose, "A Potpourri of FET Applications," *Eng. Design News (EDN)*, pp. 49–60, September 15, 1969.

10. J. D. Ryder, *Electronic Fundamentals and Applications*, 3rd ed., Prentice-Hall, Inc., Englewood Cliffs, N.J., 1964.

11. M. Schwartz, *Information Transmission, Modulation and Noise*, McGraw-Hill Book Company, Inc., New York, 1959.

12. R. Stata, "Operational Amplifiers, Parts I, II, IV," Analog Devices, Inc., Cambridge, Mass.

13. "Applications Manual for Computing Amplifiers for Modelling Measuring Manipulating & Much Else," George A. Philbrick Researches, Inc., Dedham, Mass., 1966.

14. *Communications Handbook*, Part II, Texas Instruments, Incorporated, Dallas, Tex., 1965, Chapter 6.

15. "Handbook of Operational Amplifier Applications," Burr-Brown Research Corporation, Tucson, Ariz., 1963.

16. "A Comprehensive Catalog and Guide to Operational Amplifiers," Analog Devices, Inc., Cambridge, Mass., January 1968.

17. "Power Supply Overvoltage 'Crowbars'," Hewlett-Packard Company, Palo Alto, Calif., Application Note 109, April 1969.

18. "Handbook of Operational Amplifier Active RC Networks," Burr-Brown Research Corporation, Tucson, Ariz., 1966.

19. "DC Power Supply Handbook," Hewlett-Packard, Palo Alto, Calif., Application Note 90, 1968.

20. "Applications of OEI Products," 2nd ed., Optical Electronics, Inc., Tucson, Ariz., 1968.

4

VOLTAGE DETECTORS

4-1 Introduction

The majority of signals in measurement systems ultimately appear as voltages. Since voltage cannot be seen it must be transduced into a form intelligible to the observer. One way to accomplish this is to somehow convert the time-varying voltage into a slowly varying dc voltage that is proportional to the input voltage. This dc voltage is then displayed by a meter, recorder, and so on, as discussed in Chapter 5. The converter in this case is called the voltage detector. There are four basic types of detection: peak, average, rms, and phase. In this chapter the four types of detectors are discussed along with the relative merits of each. In addition their respective readings for some common wave forms are summarized in a table. Error considerations are discussed in Section 7-11.2.

4-2 Peak Detection

A peak detector of a time-varying signal $s(t)$ simply gives the maximum value (either positive or negative) during some time interval T. Thus

$$s(t)_{\text{peak}} = \max [|s(t)|] \qquad 0 < t < T \qquad (4\text{-}1)$$

a diagram of simple peak-detector circuit is shown in Figure 4-1. The signal $s(t)$ is full-wave rectified. The capacitor C stores the rectified voltage. The resistance R is selected such that the circuit has an RC time constant which is much greater than T. At each positive peak of the applied rectified signal voltage the capacitor is charged to this peak value. As the applied voltage continues past the peak and until the next peak, the voltage across C decays slightly (as a function of the RC time constant). At the positive peak of the next cycle, however, the lost charge on C is replenished. To keep the decay time as long as possible R can be made very high through the use of a dc op amp, hence minimizing the decay between peaks. Notice that C is small since its charge time should be as rapid as possible. The meter, which itself is a mechanical averaging device, records the charge on the capacitor. This method of determining the peak value of signals is satisfactory for periodic

161

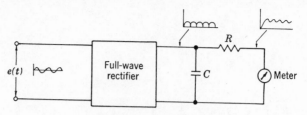

Figure 4-1 Peak detector.

or quasi-periodic wave forms. However, it is unsuited for determining the peak value of a single pulse or a pulse train with a low duty cycle (ratio of pulse duration to period). An improvement of this peak circuit whereby the peaks of single pulses of short duration can be recorded is discussed in Section 7-13.

Let us now determine the peak value of several common wave forms.

1. Sine Wave. In this case $s(t) = A \sin \omega t$, and hence

$$s(t)_{\text{peak}} = A \tag{4-2}$$

2. Square Wave. Recalling (1-9) we find that

$$s(t)_{\text{peak}} = A \tag{4-3}$$

3. Sine Wave Plus Third Harmonic. In this case $s(t) = A[\sin \omega t + b \sin (3\omega t + \varphi)]$, where $b \leq 1$. It is seen that the peak value will vary anywhere between $A(1 \pm b)$ depending on the phase angle φ.

4. Random Noise. In this case $s(t)_{\text{peak}}$ can have any value which, of course, is completely unpredictable at any given instant of time. Hence a peak measurement of a random phenomena is of little use in this form.

5. Pulse Train. Consider a pulse of duration t_0 and period T. In general, this pulse train can be expressed as

$$\begin{aligned} s(t) &= A & 0 < t < t_0 \\ s(t) &= -\alpha A & t_0 < t < T \end{aligned} \tag{4-4}$$

where $0 \leq \alpha \leq 1$. If the ac pulse train is superimposed on a dc voltage, A_{dc}, such that $A_{\text{dc}} = \alpha A$, then (4-4) becomes

$$\begin{aligned} s(t) &= A(1 + \alpha) & 0 < t < t_0 \\ s(t) &= 0 & t_0 < t < T \end{aligned} \tag{4-5}$$

If the peak detector has a blocking capacitor such that the dc component is not part of the signal, then

$$s(t)_{\text{peak}} = A \tag{4-6}$$

When the dc component is there, $s(t)_{\text{peak}} = A(1 + \alpha)$.

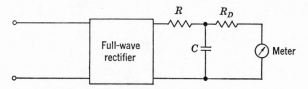

Figure 4-2   Rectified average detector.

From these examples it is seen that the peak value gives little information unless the wave form is known a priori.

4-3   Average Detection

The average or mean value of a time-varying signal $s(t)$ is, from (1-2a), with $n = 0$,

$$\langle s(t) \rangle_T = \frac{1}{T} \int_0^T s(t)\, dt \tag{4-7}$$

However, for very obvious reasons (4-7) is not a very good relation to use. Hence most average detectors measure the mean absolute or rectified value of $s(t)$, which is

$$\langle |s(t)| \rangle_T = \frac{1}{T} \int_0^T |s(t)|\, dt \tag{4-8}$$

A diagram of a simple rectified average detector circuit is shown in Figure 4-2. It is seen that R and C form a low-pass filter which in this case is being used as an integrator (recall section 2-3.2). Hence the low-pass filter continuously smooths the signal fluctuations to produce, after three or four RC time constants have elapsed, a continuous estimate of the rectified average value of the signal. The resistor R_D applies a continuous discard of previous information stored in the capacitor; thus the output is the time average of the signal level observed during the previous time constant RC. It should be mentioned that the RC averaging technique is simply an inexpensive implementation of true integration. A true integrating circuit would be an operational amplifier with a capacitor in its feedback circuit as shown in Section 3-11.7.

Let us now determine the rectified average value of several common waveforms.

1. Sine Wave.   In this case

$$\langle |s(t)| \rangle_T = \frac{2}{T} \int_0^{T/2} A \sin \frac{2\pi t}{T}\, dt = \frac{2A}{\pi} = 0.636A \tag{4-9}$$

2. Square Wave.   From Section 4-2.2

$$\langle |s(t)|\rangle_T = \frac{2}{T}\int_0^{T/2} A\,dt = A \tag{4-10}$$

Notice that this is the same value that was obtained by peak detection.

3. Sine Wave Plus Third Harmonic.   To consider a general case, let $s(t)$ be

$$s(t) = A\sin\omega t \pm Ab\sin 3\omega t \tag{4-11}$$

where b is the relative amplitude of the third harmonic to the amplitude of the fundamental and $b \le 1$. In this case (4-8) yields

$$\langle |s(t)|\rangle_T = \frac{2A}{T}\int_0^{T/2}\sin\frac{2\pi t}{T}\,dt \pm \frac{2bA}{T}\int_0^{T/6}\sin\frac{6\pi t}{T}\,dt = \frac{2A}{\pi}\left(1 \pm \frac{b}{3}\right) \tag{4-12}$$

It should be noted that the plus sign in (4-11) is the case when the third harmonic is in phase with the fundamental frequency and the minus sign is the case when it is 180° out of phase with the fundamental frequency. Comparing this result with that obtained using peak detection it is seen that both methods of detection are very sensitive to the amplitude and phase of the odd harmonics in the signal. The results of (4-12) can be generalized to the nth odd harmonic to yield the following. If $s(t) = A\sin(\omega t)\pm Ab\sin(n\omega t)$, n is odd, then

$$\langle |s(t)|\rangle_T = \frac{2A}{\pi}\left(1 \pm \frac{b}{n}\right) \qquad n = 1\ 3\ 5\ \ldots \tag{4-13}$$

It has been shown* that in the case when n is even and in phase with the fundamental, the harmonics do not affect the rectified average detector for $b \le 1.0$. It is thus seen that the rectified average detection of signals with harmonic content is very dependent on the amplitude and phase relation of the harmonics and hence raises questions as to its usefulness in measuring certain complex signals.

4. Random Noise.   Consider random noise of zero mean and variance σ_0^2 whose amplitude probability density is given by (1-94). The rectified average value of this signal is simply twice the average value of the positive amplitudes only. Thus from (1-90)

$$\langle |s(t)|\rangle = \frac{2}{\sigma_0\sqrt{2\pi}}\int_0^\infty xe^{-x^2/2\sigma_0^2}\,dx = \sqrt{\frac{2}{\pi}}\,\sigma_0 = 0.798\sigma_0 \tag{4-14}$$

5. Pulse Train.   From (4-4) and (4-5) it is seen that the rectified average value for the ac signal only is

$$\langle |s(t)|\rangle_T = A[\alpha + D(1 - \alpha)] \tag{4-15}$$

* See ref. 3.

where $D = t_0/T$ is the duty cycle of the pulse train. When $\alpha = 1$, (4-15) equals that for a square wave irrespective of the duty cycle. When $\alpha = 0$ the rectified average value is proportional to the duty cycle; that is,

$$\langle |s(t)| \rangle_T = AD \tag{4-16}$$

If the dc component is not zero (4-5) and (4-8) give

$$\langle |s(t)| \rangle_T = A(1 + \alpha)D \tag{4-17}$$

It is thus seen that rectified average detection is an improvement over peak detection. This improvement is essentially due to the fact that the rectified average value depends on the entire wave form and not, as in the case of peak detection, on only one or two points of the wave form.

4-4 Rms Detection

The rms value of a time-varying signal $s(t)$ is defined as

$$s_{\mathrm{rms}} = \langle s^2(t) \rangle_T^{\frac{1}{2}} = \left[\frac{1}{T} \int_0^T s^2(t)\, dt \right]^{\frac{1}{2}} \tag{4-18}$$

From (1-8) we see that s_{rms}^2 is the average power in a periodic signal. If the signal is aperiodic, from (1-54) we see that Ts_{rms}^2 is the total energy in an aperiodic signal where T is now the duration of the aperiodic signal. For an ergodic random signal $s_{\mathrm{rms}}^2 = R_{ss}(0)$, the autocorrelation of $s(t)$ at $\tau = 0$, which from (1-116) and (1-117) is equal to the autocovariance when the mean value of $s(t)$ is zero. Thus one can conclude that of the three types of detection introduced, rms has the strongest theoretical basis and consequently is the most meaningful measurement of all in most applications. There are essentially two different ways in which one can obtain the rms value. One way is to actually perform the squaring of the input voltage; the second is to convert the signal into heat. Before discussing how these techniques work we introduce a term called the *crest factor*, F_c, defined as the ratio of the peak value of the voltage to its rms value. Thus

$$F_c = \frac{s(t)_{\mathrm{peak}}}{s_{\mathrm{rms}}} \tag{4-19}$$

This term is one of the most important quantities used to describe the capability of an rms detector and can be thought of as the measure of the upper limit of the dynamic range of the circuit.

The first type of rms detector is shown in Figure 4-3a. An idealized version of the detector is shown in a. In b a practical implementation of the circuit is shown. The important part of the circuit is the squaring circuit. This normally is done in the very same manner as shown in Section 3-12.3. However, in Section 3-12.3 the bias to the diodes or transistors was fixed. For purposes here the bias is made variable by feeding back the voltage on the integrating

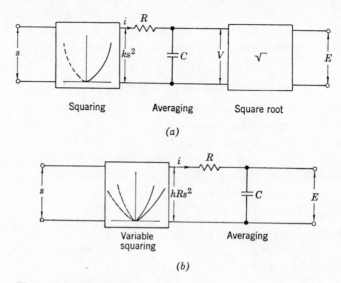

Figure 4-3 (a) Idealized rms detector and (b) practical rms detection using feedback.

capacitor C. The advantage of this variable property of the squaring circuit is that the meter scale will be almost linear and the relative error for a given wave form will be almost constant over the entire scale. The square root in this case is, therefore, not done by the meter.

In the ideal detector, however, the square root is done with the meter. In this circuit we notice that the voltage, V, on the capacitor, C, is governed by the equation

$$ks^2 - V = iR = RC\frac{dV}{dt} \tag{4-20a}$$

The output of the meter should be E. Hence if we make the substitution $E = \sqrt{V/k}$. (4-20a) becomes

$$s^2 - E^2 = RC\frac{dE^2}{dt} \tag{4-20b}$$

On the other hand, the voltage on C for the actual detector (using feedback) is of the form

$$hRs^2 - V^2 = RCV\frac{dV}{dt} \tag{4-21a}$$

where h is a constant. Upon making the substitution $V = E\sqrt{hR}$ we get

$$s^2 - E^2 = RCE\frac{dE}{dt} = \frac{RC}{2}\frac{dE^2}{dt} \tag{4-21b}$$

which is the same as (4-20b) except that the RC term differs by a factor of 2. Hence upon suitable choice of the RC product in (4-21b), the circuit shown in Figure 4-3*b* will perform exactly as that in Figure 4-3*a*. Notice that this type of detector permits one to vary the value of the integration time, RC, by varying C. Although not shown in the figures, the voltage, V, according to (4-18), must also be attenuated in amplitude by an amount $1/RC$. This is either done by calibrating the meter when RC is fixed or by attenuating the voltage with a simple resistive attenuator when RC is variable.

The second technique takes advantage of the relationship of signal energy to heat. If the signal voltage is fed into a very low resistance element (i.e., a heating filament), the element will heat up in some proportion to the voltage impressed across it. Thus the amount of heat generated by the element is directly related to the energy, $s^2(t)\,dt$, and after suitable averaging and scaling, the rms value of the signal. The heat generated is monitored by a thermo-couple which has a fixed but known averaging time and whose output is calibrated to give the rms value in volts of the input signal. One form of this type of rms detector is shown in Figure 4-4. The two thermocouples are matched for physical and mechanical properties and then mounted in the same thermal environment to overcome the nonlinear behavior of thermo-couples. Hence the nonlinear effects of the measuring thermocouple are canceled by similar nonlinear operations of the second (matched) thermo-couple. This second thermocouple is in a feedback loop. The thermocouple elements form a bridge which is unbalanced initially by the signal current present in the heater element in the ac amplifier output. The thermocouple unbalance is amplified by the dc amplifier and fed back to the heater element, reestablishing bridge balance. The dc current flowing in one heater element is then proportional to the rms value of the ac signal current in the other element and is read out directly on the meter. This technique has the advan-tage over the electronic detector previously described in that it can usually

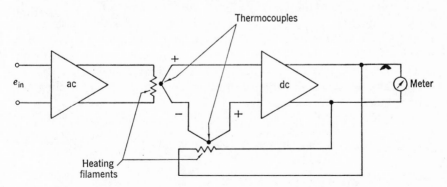

Figure 4-4 Rms detector employing thermocouples.

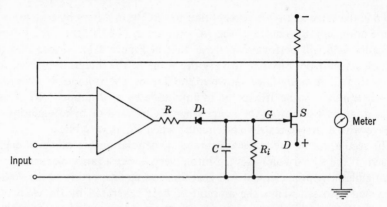

Figure 4-5 Rms detector for the measurement of impulsive-type signals.

handle high crest factors (10:1 full scale). However, it does have the disadvantage in certain applications in that there is no way to vary the averaging time of the circuit since the averaging time is a property of the thermocouple which is fixed once the thermocouple has been selected.

Before leaving the discussion of rms detectors it is worthwhile to examine a modification of the detector shown in Fig. 4-3b to measure the rms value of impulses. This measurement is finding wide use in determining acoustic noise levels from various sources, such as passing automobiles and trucks, aircraft flyover, and machinery. A typical circuit is shown in Figure 4-5. The integration is performed by RC. However, because of the diode D_1, the capacitor C cannot discharge through R when the input signal decreases from its maximum value. In order to keep this voltage level at its peak value, which is attained at time $t = RC$, the field effects transistor (FET) is used. The FET has, typically, an extremely high-input impedance $R_i (> 1000 \text{ M}\Omega)$. Because of the diode, the discharge time constant is now $R_i C$ which is easily made to be of the order of magnitude of 30 min before the signal decreases 1 dB from its peak amplitude. This storage time could be made longer if it were not for the inevitable leakage of the various components. It should also be pointed out that the input impedance of the FET is highly temperature dependent and decreases with increasing temperature, thus affecting the storage time.

Let us now determine the rms value of several common wave forms.

1. Sine Wave. In this case (4-18) yields

$$s(t)_{\text{rms}} = \left[\frac{A^2}{T} \int_0^T \sin^2 \frac{2\pi t}{T} \, dt \right]^{1/2} = \frac{A}{\sqrt{2}} = 0.707A \qquad (4\text{-}22)$$

2. Square Wave. From example 2, in Section 4-3 we find

$$s(t)_{\text{rms}} = \left[\frac{A^2}{T} \int_0^{T/2} dt + \frac{A^2}{T} \int_{T/2}^T dt \right]^{\frac{1}{2}} = A \qquad (4\text{-}23)$$

The rms value of a square wave is the same as its peak and rectified average value.

3. Sine Wave Plus Third Harmonic. Using (4-11) and (4-18) gives

$$s(t)_{\text{rms}} = \left[\frac{A^2}{T} \int_0^T \left[\sin \frac{2\pi t}{T} \pm b \sin \frac{6\pi t}{T} \right]^2 dt \right]^{\frac{1}{2}} = A \left(\frac{1 + b^2}{2} \right)^{\frac{1}{2}} \qquad (4\text{-}24)$$

It is seen that this result is completely independent of any phase relationship of its harmonics [recall (1-8)].

4. Random Noise. Using (1-90) and (1-94) it is seen that

$$s(t)_{\text{rms}} = \left[\frac{1}{\sigma_0 \sqrt{2\pi}} \int_{-\infty}^{\infty} x^2 e^{-x^2/2\sigma_0^2} dx \right]^{\frac{1}{2}} = \sigma_0 \qquad (4\text{-}25)$$

which, of course, is the definition of σ_0 for zero mean.

5. Pulse Train. From (4-5) and (4-18) it is seen that the rms value of the signal, including the dc component, is

$$s(t)_{\text{rms}} = A(1 + \alpha)\sqrt{D} \qquad (4\text{-}26)$$

When the dc component is blocked by a transformer or capacitor in series, (4-4) and (4-18) give

$$s(t)_{\text{rms}} = A[D(1 - \alpha^2) + \alpha^2]^{\frac{1}{2}} \qquad (4\text{-}27)$$

The crest factor F_c can now be computed using (4-4) and (4-5). For the case where the dc component is included we get

$$F_{c_{(dc)}} = \frac{1}{\sqrt{D}} \qquad (4\text{-}28a)$$

and when only the ac component is present

$$F_{c_{(dc)}} = [D(1 - \alpha^2) + \alpha^2]^{-\frac{1}{2}} \qquad (4\text{-}28b)$$

There is another limitation which one must consider when making an rms measurement and that is the bandwidth of the detector. From (1-8) it is known that the rms value is the square root of the sum of the squares of amplitudes of the harmonic contribution to a given periodic signal. From (1-19) it is seen that for the case of the dc pulse train we have

$$s(t)_{\text{rms}} = 2(1 + \alpha)D \left[1 + 2 \sum_{k=1}^{\infty} \left(\frac{\sin k\pi D}{k\pi D} \right)^2 \right]^{\frac{1}{2}} \qquad (4\text{-}29)$$

If the frequency response of the measuring device is less than the spread of the harmonics, some will be excluded from the measurement and an error introduced. It is a relatively easy matter to show that if the bandwidth of the system is $B = 1/t_0$ the loss in amplitude due to the exclusion of the higher harmonics yields a 5% error; if $B = 5/t_0$ the error is 1% and if $B = 10/t_0$ the error reduces to 0.5%. A bandwidth of $50/t_0$ is required to reduce the error to 0.1%.

It is seen that the rms measurement does not have the drawbacks of peak and rectified average detection; that is, it is not sensitive to harmonic content as is the average detector. It could be argued that the rectified average detector's reading can be corrected if the harmonic content is known. While the amplitude of each harmonic component may be measured with reasonable accuracy, instruments for measuring accurately phase relative to the fundamental are not available. Perhaps a more important reason for using rms is brought out by the simple fact that most random signals measured in actual practice cannot be described by (1-94). Thus the factor 0.798 appearing in (4-14) may not be determinable to any degree of accuracy or certainty. In this case then the average rectified value is of little use. In addition, the rms value is from theoretical considerations well founded, and related to meaningful quantities used to describe signal content; namely, energy and power.

4-5 Summary of Results of Peak, Rectified Average, and RMS Detection

The results of the five examples given in each of the sections on the peak, average, and rms detectors are summarized in Table 4-1. Since it is common

Table 4.1

Wave Form	Rms Meter Indicates	Rectified Average Meter Indicates (Calibrated in Rms)	Peak Meter Indicates (Calibrated in Rms)	Error (dB) $[20 \log_{10} (E/E_{rms})]$	
				Average	Peak
Sine wave	0.707	0.707	0.707	0	0
Sine wave Plus 100% Third harmonic					
In phase	1.000	0.944	1.09	−0.50	0.748
Out of phase	1.000	0.472	0.382	−6.52	−8.36
Square wave	1.000	1.111	0.707	+0.91	−3.00
Gaussian noise	σ_0	$0.887\sigma_0$	—	−1.04	—
Pulse train ($\alpha = 0$)					
$D = 0.1$	0.318	0.111	0.707	−9.14	+6.10
$D = 0.01$	0.10	0.011	0.707	−19.17	+16.99

practice for many commercial detectors to use scales that are calibrated to read rms voltage of a *sine wave*, Table 4-1 depicts what the peak, average, and rms meter will read when calibrated in rms for a sine wave, and what the error in dB is compared to the true rms reading. This table is obtained directly from the previous examples by setting $A = 1.0$.

Application 4-1   Determination of the Type of Unknown Detector

It is seen from Table 4.1 that one can easily determine whether a detector is rms, rectified average, or peak by performing the following simple test. Give as a signal input to the detector a square wave and then a sine wave both of whose peak amplitudes are equal. Record the meter readings. If it is a true rms detector the sine wave reading will be 0.707 of the square wave reading. If it is a rectified average detector calibrated in rms, the sine wave reading will be 0.785 of the square wave. If the detector is a peak detector both readings will be equal.

4-6   Phase Detector (Phase-Lock Amplifier)

A phase-lock amplifier is a device which can be used to determine the amplitude of extremely weak periodic signals buried in noise. Signals as weak as 10 nV can be detected with this type of device. Referring to Figure 4-6 it is seen that the input signal is amplified approximately 80 dB with cascaded amplifiers, the last stage of which is a variable-frequency tuned amplifier. This tuned amplifier is simply an active filter with gain, wherein the properties of the filter are varied to obtain a continuously variable center frequency. The purpose of this tuned amplifier is to partially filter the input signal thus giving the electronics following a better dynamic range. The output of the tuned amplifier is fed into a multiplier. The other signal to the multiplier is the reference signal. The reference signal first passes through

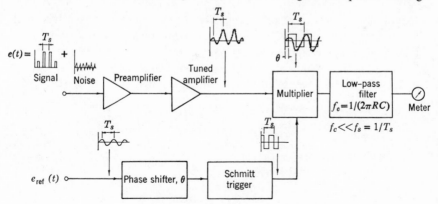

Figure 4-6   Phase-lock amplifier.

a phase-adjusting (shifting) circuit and then into a Schmitt trigger (see Sections 3-11.6 and 5-6.2 for a discussion of this device) set to trigger on the zero crossings of the reference signal. The output of the Schmitt trigger is a square wave of period T_s, the period of the reference signal, having a variable phase shift θ with respect to the reference signal.

The output of the multiplier is the product of the phase-shifted square wave and the input signal. Due to the filtering by the tuned amplifier, its output will be primarily the first harmonic of the periodic signal $e(t)$ plus any noise that is present in the bandwidth. Ignoring for the moment any additive noise in the input signal, we see from (2-62) and (1-11) that when the reference signal and the input signal have exactly the same frequency the output of the multiplier will be of the form

$$g(t) = A_1 \cos \theta + \sum \text{ higher harmonics} \qquad (4\text{-}30)$$

The output $g(t)$ is then passed through a low-pass RC filter. From Section 2-3.2 we recall that the effective bandwidth is $B_e = 1/(4RC)$ for a single-stage filter and $B_e = 1/(8RC)$ for a double stage. Hence if a two-stage filter is used with, say, $RC = 10$ sec, then the original signal is filtered with a bandwidth of $B_e = 0.0125$ Hz. If noise is added to the periodic signal, the noise bandwidth has been reduced to 0.0125 Hz. Obviously, as the RC time is increased the noise bandwidth is reduced proportionately thus increasing the signal-to-noise ratio. Furthermore, the output of the periodic signal can be maximized by adjustment of the phase-shift angle θ. Since the phase at a fixed frequency will remain constant for a given system the amplifier is said to be "locked" onto the signal. Small amplitude variations of the signal itself will affect the output minimally due to the slow response time of the low-pass filter.

Application 4-2   Measurement of Harmonic Distortion ($\leq 10\%$)

To measure percentage harmonic distortion we use the twin-tee notch filter discussed in Section 2-4.6 and a true rms detector. We first measure the rms value of the signal using the rms meter. This gives a value for the denominator of (1-25) [recall (1-6)]. We now pass the signal through the notch filter, which is centered at the fundamental frequency of the signal,

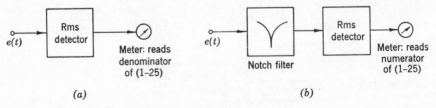

(a)   (b)

Figure 4-7   Measurement of harmonic distortion less than 10%.

and read the output with the rms meter. See Figure 4-7. Since the fundamental frequency has been virtually eliminated we have a value for the numerator of (1-25). Taking the ratio of these two values and multiplying by 100 yields the percentage of total harmonic distortion.

It should be pointed out that the minimum percentage distortion that can be measured by this method is determined by the amount of attenuation of the notch filter. If the notch filter decreases the amplitude of the fundamental frequency to a value α, then from the numerator of (1-25) we see that

$$\alpha^2 \ll \sum_{j=2}^{N} a_j^2 \qquad N > 2$$

in order that this technique be valid. If $\alpha = 10^{-3}$ (60-dB attenuation) then the technique is valid for harmonic distortion as small as 1%.

Application 4-3   Measurement of Weak Radiant Energy

Consider the case wherein information is in the form of radiant energy. This radiant energy is detected by a photosensitive device. The ambient level of the background radiant energy is such that the signal of interest is buried in the background noise. Assuming that the energy source contains information which is very slowly varying we employ a mechanical chopper to translate this slowly varying information to a frequency f_0. (A good choice of f_0 is around 100 Hz, which is mechanically feasible to obtain and lies between the 60 Hz line voltage and is first harmonic, 120 Hz). See Figure 4-8. The mechanical chopper transmits a pulse train whose period coincides with the mechanical chopping of the radiant energy. Any phase differences can be adjusted for with the phase-shifting device in the phase-lock amplifier.

The effect of this mechanically induced frequency translation of the signal information is to remove it from the $1/f$ (Johnson) noise region of the amplifiers and additionally to create a "reference" frequency (signal) with which one can use to help distinguish the signal from the noise. Thus the reference frequency is fed into the reference channel of the amplifier and the

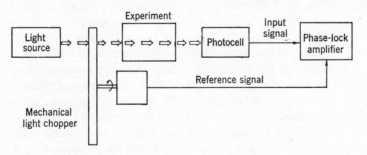

Figure 4-8   Measurement of weak radiant energy.

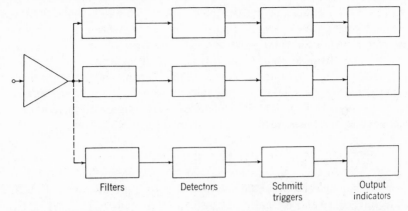

Filters   Detectors   Schmitt   Output
                                   triggers   indicators

Figure 4-9   Quality control testing employing detection of acoustic or vibratory properties of specimen.

output from the photodetector to the input channel. If the original radiant energy had a bandwidth $\Delta f \ll 1$ centered around zero frequency, the output of the tuned amplifier will have a bandwidth of 2 Δf centered about f_0. The output of the phase detector will be a very slowly varying dc signal whose amplitude is the average amplitude of the original signal of bandwidth Δf. It has been tacitly assumed in this illustration that $\Delta f \ll B_e$, wherein B_e is the effective bandwidth of the integrating circuit of the amplifier. If this is not the case the output of the lock-in amplifier will be varying in time at the approximate rate of $1/B_e$.

Application 4-4   Quality Control Testing

From previously conducted tests it has been found that certain devices exhibit high sound or vibration levels in certain frequency levels when their operation becomes defective. This fact can be used to perform quality control tests on finished products off an assembly line and a typical system is shown in Figure 4-9. When sound pressure level is to be measured the item is placed in an enclosure designed for sound measurements and then energized. When vibration levels are to be measured an accelerometer is placed on the item and the item is energized. The accelerometer or microphone is connected to an impedance-matching amplifier (see Applications 3-2 and 3-3) and then to a voltage amplifier. Connected to the voltage amplifier is a set of parallel filters of preselected bandwidth and center frequency followed by detectors and comparators. The level of the comparators have been preset to some voltage level deemed to indicate that the item is defective. When the vibration or the sound pressure levels exceed the preset level an alarm is triggered and the item is removed from the assembly line.

References

1. J. J. Davidson, "Average vs RMS Meters for Measuring Noise," *IRE Trans. Audio*, July–August 1961, pp. 108–111.
2. J. C. Fisher, "Lock in the Devil, Educe Him, or Take Him for the Last Ride in a Boxcar?" *TEK TALK*, Vol. 6, No. 1, Princeton Applied Research Corporation, Princeton, N.J.
3. J. G. Nordahl, "Evaluating Performance of AC-DC Converters," *Electron. Instrum. Dig.*, **3,** No. 11 (November 1967), pp. 7–12.
4. E. Uiga, "How to Measure the RMS Value of Pulse Train," *EEE*, August 1968, pp. 102–106.
5. C. G. Wahrmann, "A True RMS Instrument," *B & K Tech. Rev.*, No. 3, July 1958, pp. 9–21.
6. C. G. Wahrmann, "Methods of Checking the RMS Properties of RMS Instruments," *B & K Tech. Rev.*, No. 1, 1963.
7. C. G. Wahrmann, "Impulse Noise Measurements," *B & K Tech. Rev.*, No. 1, 1969, pp. 7–18.

5

RECORDERS

5-1   Introduction

Recorders are used to provide a record of the amplitude variations of a signal as a function of time. The actual wave form may or may not be displayed depending upon whether or not there is a detector in the recorder. If a detector is included only some amplitude property of the input signal is recorded. One exception to wave form or amplitude recording is the magnetic tape recorder where the recording can be used to examine the event either in its true time or in a scaled time. This chapter discusses the most common types of recorders, their limitations, and some of their applications.

5-2   Oscillograph Recorder

The oscillograph recorder, or direct-writing recorder, records either the actual wave form of the signal as a function of time or the detected value (recall Chapter 4). A writing mechanism passes over the recording paper which may be ink or light sensitive. For ink systems the writing mechanism is a pen or scribe. If the paper is light sensitive the mechanism is a mirror. The writing mechanism is called a galvanometer and is shown in Figure 5-1 with its equivalent circuit. Viscous damping, c, may or may not be present. The galvanometer is an electromechanical device which transforms electrical energy into mechanical rotation. As shown in Figure 5-1 an input voltage e_g from a signal source with output impedance R_g causes a current to flow in the coil. The coil is surrounded by a permanent magnetic field; thus the coil is impressed with an electromagnetic force which causes a torque. This torque tends to rotate the coil an amount θ_0 until it is balanced by the restoring torque of the torsion springs.

Writing equations for the current, i_g, impressed on the coil, it is found that

$$i_g(R_c + R_g) + L_c \frac{di_e}{dt} - e_g = -HNLb \frac{d\theta_0}{dt} \qquad (5\text{-}1)$$

where the right-hand side of (5-1) is the back emf of the coil acting as a generator, H is the magnetic flux density, N is the number of turns on the

177

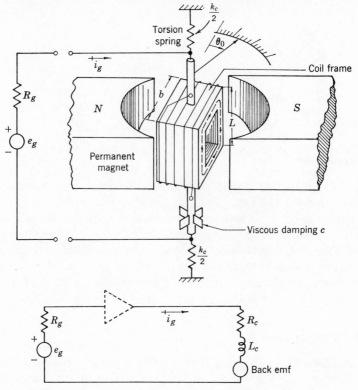

Figure 5-1 A galvanometer and its electrical equivalent circuit.

coil, L is the length of the coil, b is the depth of the coil, L_c is the inductance of the coil, and R_c is the resistance of the coil. Within the useful operating frequency range of the galvanometer the effect of the inductance, L_c, is negligible and, therefore, is neglected.

Summing torques in the mechanical system it is found that

$$HNLbi_g - \frac{(HLb)^2}{R_f}\frac{d\theta_0}{dt} - k_c\theta_0 - c\frac{d\theta_0}{dt} = J_c\frac{d^2\theta_0}{dt^2} \tag{5-2}$$

where the first term in (5-2) is the electromagnetic torque, the second term is the torque induced in the coil frame, R_f is the resistance of the coil frame, k_c is the torsional constant, J_c is the moment of inertia of the moving coil and spring, and c is the viscous damping. Solving (5-1) for i_g for the case of harmonic oscillations and substituting the result into (5-2) it is found that

$$\frac{\theta_0(\omega)}{e_g} = K\left(\frac{\omega^2}{\omega_n^2} + \frac{2\xi\omega}{\omega_n} + 1\right)^{-1} \tag{5-3}$$

where $K = HNLb/[k_c(R_g + R_c)]$, $\omega_n = \sqrt{k_c/J_c}$, and

$$\xi = \frac{c + (HLb)^2/R_f + (HNLb)^2/(R_c + R_g)}{2\sqrt{k_cJ_c}} \qquad (5\text{-}4a)$$

From (5-3) we see that the sensitivity, K, varies with R_g. We note further from (5-4a) that when the viscous damping is zero the system is still damped. The damping that remains is electromagnetic which is caused by the back emf. It can be shown* that for maximally flat response $\xi = 0.68$. In low-frequency galvanometers the electrical resistance is adequate to obtain optimum damping and no fluid damping is used. Also, such galvanometers may have the coil frame slotted so that R_f is infinite. When R_f is infinite (very large), the damping is given by

$$\xi = \frac{(HNLb)^2}{2\sqrt{k_cJ_c}(R_c + R_g)} \qquad (5\text{-}4b)$$

From (5-4b) it is apparent that the damping depends upon the generator output impedance R_g. Thus such a galvanometer can be properly damped for only one value of external resistance. When the resistance of the generator is low compared to the damping resistance required to satisfy (5-4b), a resistance must be connected in series with the generator and the galvanometer. If the generator resistance is high compared to the required damping resistance, a shunting resistor will be needed. The parallel value of the source resistance and the shunt resistor must equal the required damping resistance to obtain $\xi = 0.68$. If an amplifier is placed between the galvanometer and the generator, the optimum damping can be obtained for a wide variety of generator resistances provided only that $R_g/R_a \ll 1$, where R_a is the input resistance of the amplifier and R_g in (5-4b) equals the output resistance of the signal source.

Note from (5-3) that the upper limiting frequency depends on the natural frequency of the system. Thus for maximum high-frequency response either the spring constant must be increased or the moment of inertia must be decreased. In general the spring constant is increased and as (5-3) shows, with a loss of sensitivity. When the spring constant is increased the electromagnetic damping is inadequate, and internal viscous damping is required. In such cases the electromagnetic damping is a small percentage of the total damping and the galvanometer can be used with a variety of generator impedances. The galvanometer is used with a chart paper moving at a known constant speed. If the galvanometer is to be used with an ink writing system, an ink pen is attached to the moving coil. However, due to the added inertia this system is usually limited to 200 Hz. Another system replaces

* See Figure 6-15.

the ink pen with a mirror. When the coil turns a light beam, which is focused as a spot on a moving light-sensitive paper, exposes a trace of the wave form. In general this system is limited to 600 Hz for resistively damped galvanometers and 13,000 Hz for fluid-damped galvanometers.

An electrical signal which passes through the galvanometer coil causes the writing mechanism to rotate about an axis defined by the suspension. The physical rotation, θ, of the mirror is given by $\theta = H(\omega)i(\omega)\cos\theta$ where $H(\omega)$ is a constant in the useful frequency range of the galvanometer and $i(\omega)$ is the current to the galvanometer. The linear deflection, d, on the paper will be proportional to the distance, l_0, from the galvanometer to the actual point of writing and is given by

$$d = \frac{H(\omega)i(\omega)l_0 \sin\theta}{\theta(1 - \tan^2\theta)}$$

To account for the departure from a linear relationship for large $\theta(\theta > \pm 9°)$, either a tolerance is added to the recorder specifications, or in ink systems, calibrated paper is used.

5-3 X-Y Plotters

The X-Y plotter is a servo controlled recorder for recording one variable on one coordinate (e.g., X) and another variable on another coordinate (e.g., Y). If the plotter is equipped with a time-base generator it can also record one variable versus time. Consider the block diagram shown in Figure 5-2. The input voltage is first passed through a stepped attenuator and then through a potentiometer. The position (amount of attenuation) of the continuously variable potentiometer is determined by the position of the plotter pen. The output voltage from the potentiometer is amplified with a chopper-stabilized amplifier and compared to a reference voltage by means

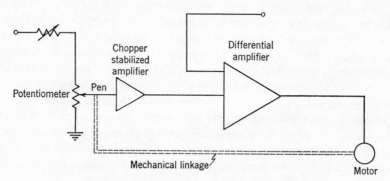

Figure 5-2 One axis of an X-Y plotter.

of a differential amplifier (see Section 3-11.5). Any difference between these two voltages is applied to a motor which drives the recording pen. The recording pen is mechanically coupled to the potentiometer. If the writing pen is positioned such that the output voltage from the potentiometer is exactly equal to the reference voltage, the pen will not move. However, when the voltage from the potentiometer is not equal to the reference voltage, a voltage will be applied to the motor which will cause the motor and pen to move in proportion to the voltage, thus obtaining a a recording.

Both the X-axis and the Y-axis operate in the same manner; however, some plotters have the provision for a time base. If a constant voltage (dc) is integrated, a ramp function in time will be generated. This ramp function can be applied to the X-axis of the recorder so that the recorder can be used to plot voltage as a function of time.

It is clear that the motor must take more time to travel full scale than half scale. For this reason the upper frequency limit of an X-Y plotter is specified in writing (or slewing) rate rather than frequency [recall (3-29)].

The same servo system can be used as a strip chart recorder by using only one axis of recording and advancing paper at a constant rate. This type of continuous recorder does not require calibrated paper to correct for the rotational to rectangular coordinates as do the galvanometer-type recorders.

5-4 Graphic Level Recorder

Level recorders function in a manner similar to a strip chart recorder, and additionally, they contain voltage detectors and variable writing speeds (detector integration times), and they can be used synchronously with other instruments. The level recorder is essentially a recording voltmeter. An input voltage is passed through an input potentiometer to a calibrated input attenuator as shown in Figure 5-3. The signal is then connected to a second input potentiometer (termed the range potentiometer) which consists of a bank of precision resistors (or a nichrome strip) on which the slider of a driving arm moves. The resistors are spaced along the length of the laminate such that either a linear or a logarithmic attenuation of the signal can be obtained. The importance and advantage of the range potentiometer will be explained subsequently.

The position of the slider on the laminate determines the range potentiometer output voltage. This output voltage is amplified and then rectified using either rms, peak, or average detection. The output of the detector is then averaged (integrated) with a low-pass RC filter. This signal is then compared to a reference voltage. If any difference exists, it is amplified and applied to the driving coils of the writing system (driving arms). The writing system is

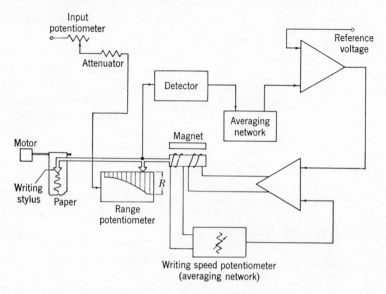

Figure 5-3 Functional diagram of a level recorder.

mechanically coupled to the writing stylus. If the slider is positioned such that the output voltage from the potentiometer is exactly equal to the reference voltage, the writing stylus will not move. However, when the voltage from the detector is not equal to the reference voltage, a voltage will be applied to the coils of the writing system which will cause the slider and writing stylus to move in proportion to the voltage, thus obtaining a recording. Note that since the difference between the reference voltage and the attenuated voltage (from the range potentiometer) is detected and amplified, the dynamic range of the system does not depend on the dynamic range of the amplifier or detector but only upon the dynamic range of the range potentiometer. However, a large dynamic range means a decrease in the resolution since the size of the chart paper remains constant.

A secondary winding on the driving coils is used to obtain a velocity dependent signal. As shown in Figure 5-3, this signal is introduced into a differential amplifier as a feedback signal. If the error signal (the difference between the signal from the detector and the reference signal) is e_s and the feedback signal is e_{fb}, then the output of the differential amplifier is

$$e_o = e_s - ke_{fb} = e_s - k\frac{de_s}{dt}$$

where k is the writing-speed potentiometer constant. Thus if the error signal changes rapidly and k is such that $e_s = ke_{fb}$ the writing stylus will not move. However if k is slightly less than e_s/e_{fb} the writing stylus will move but very slowly. Since the attenuator can be changed the writing rate (averaging time) of the level recorder can be changed.

Recall that averaging was obtained using the capacitors following the detector. Averaging is therefore obtained by both the detector and the feedback signal. The expression for the recorded level, A is (assuming rms detection is used)

$$A = \frac{1}{T_2} \int_0^{T_2} \left[\frac{1}{T_1} \int_0^{T_1} e^2(t) \, dt \right]^{1/2} dt$$

where T_1 is the effective averaging time of the capacitor circuit and T_2 is the averaging time of the writing system. It can be shown that if $T_1 > T_2$ the system is unstable. However, if T_2 is chosen to be much greater than T_1 and the frequencies of the fluctuations in the detector input level are of the order of $1/T_1$ the recorded signal level will be the "mean detected value" (i.e., "mean rms"). This quantity is of dubious physical meaning and from a theoretical point of view it is better to increase T_1 and do away with the second integration entirely. It should also be noted that (2-66) through (2-68) apply with equal validity to level recorders.

The paper drive is powered by a synchronous motor which is connected to a gear box. The gear box powers a drive shaft which is synchronized with the paper drive; thus external instruments such as an oscillator or frequency analyzer can be driven with the level recorder. The drive shaft can also be used for actuating microswitches for pulsed motors and so on.

5-5 Magnetic Tape Recording

5-5.1 Direct Record

The record head of a tape recorder is similar to a transformer with a single winding. Signal current flows in the winding, producing a magnetic flux in the core material. To perform as a record head, the core is made in the form of a closed ring with a short nonmagnetic gap. When the nonmagnetic gap is bridged by magnetic tape, the magnetic flux detours around the gap through the tape completing the magnetic path through the core material. The magnetic tape consists of a plastic ribbon on which tiny particles of magnetic material have been deposited uniformly. When the tape is moved across the record-head gap, the magnetic material (termed oxide) is subjected to a flux pattern which is proportional to the signal current in the head winding. As it leaves the head gap each particle retains the state of magnetization that was last imposed on it by the shunted gap. Thus the actual recording takes place at the trailing edge of the record-head gap. A simplified diagram of the recording head and magnetic tape is shown in Figure 5-4.

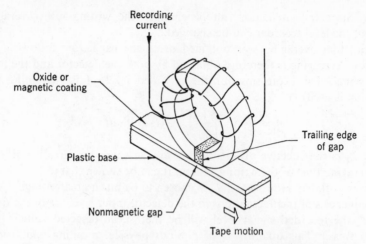

Recording
current

Oxide or
magnetic coating

Trailing edge
of gap

Plastic base

Nonmagnetic gap

Tape motion

Figure 5-4   Simplified diagram of the magnetic head and magnetic tape.

To reproduce the signal, the magnetic pattern on the tape is moved across a reproduce head. A small nonmagnetic gap in the head is bridged by the magnetic oxide of the tape. Magnetic lines of flux are shunted through the core and are proportional to the magnetic gradient of the pattern on the tape which is spanned by the gap. The induced voltage in the head winding follows the law of electromagnetic induction $e = N d\varphi/dt$, where N is the number of turns on the core and φ is the magnetic flux. It is important to note that the reproduced voltage is not proportional to the magnitude of the flux, but to its rate of change.

Suppose the signal to be recorded on the tape is a sine wave described by $A \sin(2\pi ft)$. Both the current in the record-head winding and the flux, φ, through the record-head core will be proportional to this voltage. If the tape retains this flux pattern and regenerates it in the reproduce head core, the voltage in the reproduce head winding will be

$$e_{\text{reproduce}} = k \frac{d\varphi}{dt} = 2k\pi fA \cos(2\pi ft)$$

Thus the reproduce head acts as a differentiator and the reproduced signal is actually the derivative of the recorded signal and not the signal itself. The output of the reproduce head is proportional to the signal frequency, and therefore, to maintain a constant frequency response the 6-dB/octave increase in the head output must be compensated for in the reproduce amplifier by a process termed equalization. Figure 5-5 shows how such a shaped response is applied. The decrease in output voltage at high frequency will be explained in Section 5-5.3. At some low frequency the output voltage from the reproduce head falls below the inherent noise level of the overall recording

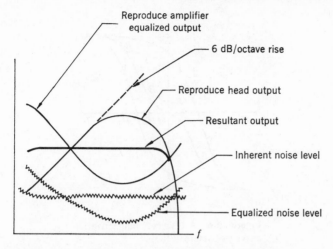

Figure 5-5 Reproduce characteristics and equalization of the reproduce head.

system. Thus there is a low-frequency limit in the direct process below which reproduction cannot be made. If the tape speed is changed the equalizers must also be changed since a different frequency weighting is required. However, no change is necessary in the record amplifier.

An audio recorder may be considered a special case of a direct recorder. The characteristics of the record and reproduce amplifiers are modified to conform to the particular characteristics of speech and music. It has been found that the spectral content in speech and music signals is not uniformly distributed over the range of audio frequencies. For this reason preequalization circuits are incorporated in the record amplifier which amplify selected portions of the frequency spectrum (extreme low and high frequencies). By increasing the magnitude of the low and high frequencies, it is possible to approach a constant flux density on the tape at all frequencies. In this way the signal-to-noise ratio is improved. During the reproduce process the output must be equalized in an inverse fashion.

The problem in using an audio recorder for instrumentation recording is that most signals do not have the peculiar spectral density of speech or music. The result is that the weighting in the record amplifier could result in serious distortion of the high and low frequencies. This could only be overcome by reducing the recording level by a considerable amount at the cost of decreasing the signal-to-noise ratio.

5-5.2 Bias

Like other magnetic materials, the particles on the tape exhibit a very nonlinear characteristic when exposed to a magnetizing force. A typical

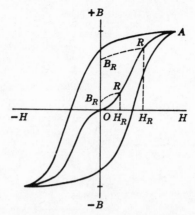

Figure 5-6 A typical magnetization curve or hysteresis loop.

magnetization curve, or hysteresis loop, is shown in Figure 5-6. In Figure 5-6, H is the magnetizing force and is determined by the number of turns and the current is the record-head winding. The quantity B is the induced magnetization on the tape.

As a demagnetized particle in the tape approaches the record-head gap it carries no residual magnetism (the origin in Figure 5-6). Assuming that a cycle of the recorded signal along the tape is very long compared to the gap length, the particle will pass through an essentially constant magnetizing force created by the recording current. Such a force, H_R, will carry the magnetization of the particle up the curve OA to point R, at the center of the gap. As the particle leaves the gap, H falls to zero, but the magnetization of the particle will follow a minor hysteresis loop, $R - B_R$, retaining a residual magnetization B_R. The transfer characteristic of this process is shown graphically in Figure 5-7 and the inherent nonlinearity is readily apparent. High distortion in the reproduced signal results unless some corrective action is taken.

There are two fairly linear segments in the transfer characteristic curve, one on each side of the origin with their centers about half way to saturation. If the recording can be confined to one (or both) of these straight sections, low distortion can be realized. If both linear sections of the curve are to be used some means of rapidly switching from one to the other must be devised. This is exactly what high-frequency ac bias does. The input signal is mixed with a high-frequency signal termed the bias signal. If the amplitude of this signal is correctly chosen, recording will take place in the linear range of the magnetic material. Since the bias itself is not reproducible because of its high frequency, its switching function is not detectable and the gap between

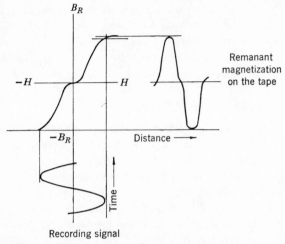

Figure 5-7   Head-to-tape transfer characteristics with no bias.

the two linear sections disappears. Figure 5-8 shows graphically how a low-distortion magnetic signal is recorded.

Since the bias and the signal are linearly mixed together no new frequency components are produced. The proper amplitude for the bias is dependent upon the exact transfer characteristics of the tape and should be adjusted to

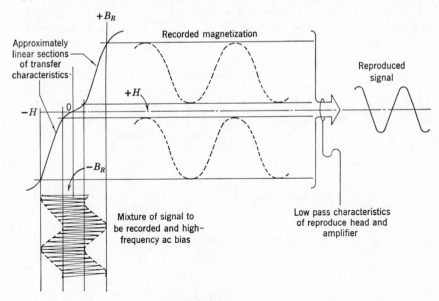

Figure 5-8   Graphical representation of ac bias recording.

reach from center to center of the linear regions. Too much bias will greatly reduce the high-frequency response, while inadequate bias will cause increased distortion of the lower frequencies. In general, bias frequencies are not critical, but should be at least 3.5 times the highest frequency to be recorded to minimize interaction with harmonics of the signal.

5-5.3 High-Frequency Response

If a sine wave is recorded, the magnetic intensity of the recorded track will vary sinusoidally. The distance along the tape required to record a complete cycle is termed the recorded wavelength, λ, and is directly proportional to tape speed and inversely proportional to signal frequency. For example, consider a tape recorder with an upper limiting frequency of 1.2 MHz at 120 in./sec (ips) recording speed. Such a signal has a wavelength of 0.0001 in. which is termed the limit of the machine's resolution. It may also be said that the machine is capable of a packing density of 10,000 Hz/in. Both packing density and resolution can be used to describe a recorder's response independent of tape speed and thus are more definitive of a recorder's capability than just a frequency specification at a given speed.

As shown in Figure 5-9, the reproduce head output increases with frequency up to a point and then decreases rapidly to zero. The decrease is primarily the result of gap effect and occurs as the recorded wavelength, λ, becomes shorter and eventually equals the reproduce gap dimension itself. At this point there is no magnetic gradient spanned by the gap, and thus no output voltage. This is the most serious, single restriction on a tape recorder's high-frequency response. It would appear that the gap width could be made

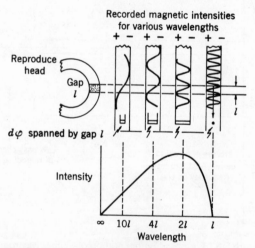

Figure 5-9 Graphical representation of the gap effect.

Table 5-1   Single-Carrier and Wide-Band FM Record Parameters

Tape Speed (ips)			Carrier Center Frequency (Hz)	Carrier Deviation Limits		Modulation Frequency (Hz)	Response at Band Limits (dB)[a]
Low Band	Inter-mediate Band	Wide Band		Carrier Plus Deviation (Hz)	Carrier Minus Deviation (Hz)		
$1\frac{7}{8}$			1,688	2,363	1,012	Dc 313	±1
$3\frac{3}{4}$	$1\frac{7}{8}$		3,375	4,725	2,025	Dc 625	±1
$7\frac{1}{2}$	$3\frac{3}{4}$		6,750	9,450	4,050	Dc 1,250	±1
15	$7\frac{1}{2}$	$3\frac{3}{4}$	13,500	18,900	8,100	Dc 2,500	±1
30	15	$7\frac{1}{2}$	27,000	37,800	16,200	Dc 5,000	±1
60	30	15	54,000	75,600	32,400	Dc 10,000	±1
	60	30	108,000	151,200	64,800	Dc 20,000	±1
	120	60	216,000	302,400	129,600	Dc 40,000	±1
		120	432,000	604,800	259,200	Dc 80,000	±1
		Wideband Group 2[a]					
		120	900,000	1,170,000	630,000	Dc 400,000	±3
		60	450,000	585,000	315,000	Dc 200,000	±3
		30	225,000	292,500	157,500	Dc 100,000	±3
		15	112,500	146,250	78,750	Dc 50,000	±3
		$7\frac{1}{2}$	56,250	73,125	39,375	Dc 25,000	±3
		$3\frac{3}{4}$	28,125	36,562	19,688	Dc 12,500	±3

[a] Frequency response referred to 1-kHz output for FM channels 13.5 kHz and above and 100 Hz for channels below 13.5 kHz. The second group of wide-band FM carrier frequencies is primarily for use with predetection recorders where one or more analog channels are also required.

smaller so that the recorder could record higher frequencies. However, as gap size is reduced the dynamic range will also decrease. Hence to have the same dynamic range but increased frequency response, the tape speed must be increased. For most recorders, tape speed and upper limiting frequency correspond to standards set by the inter-range instrumentation group (IRIG). These standards are listed* in Table 5-1.

Magnetic tape is manufactured in such a way that the magnetic particles are ground to a small size and applied to the surface of the tape in as uniform a coating as possible. Surface defects of various types occur. The most serious of these are clusters of oxide particles which form along the surface of the tape. As these particles pass across the head, they cause the tape to be lifted from the head and result in a drop in signal level. A similar drop in signal level will occur if a particle of dust or dirt is permitted on the surface of an otherwise perfect tape. It can be shown that the loss in amplitude L, expressed in decibels, is given by $L = 54D/\lambda$, where D is the separation of the tape from the head, and λ is the recorded wavelength on tape. Thus the effect is most serious at the shorter recorded wavelengths (high frequency signals) which approach the size of the nodules or foreign particles. The loss of the signal in this manner is commonly called "drop out."

* See ref. 8.

5-5.4   Frequency-Modulated (FM) Recording

To extend the lower limiting frequency of a tape system, FM recording is used. Data recording using a frequency-modulated carrier is accomplished by deviating the carrier frequency in response to the amplitude of an input signal. Thus a dc signal of one polarity increases the carrier frequency by some amount and the opposite polarity decreases it. An ac signal alternately increases and decreases the carrier above and below its center frequency at a rate equal to the input signal frequency.

Mathematically an FM signal can be expressed as

$$f(t) = A_c \cos (2\pi f_c t + X \cos 2\pi f_1 t) \tag{5-5}$$

where f_c is the carrier frequency, f_1 is the signal frequency, and X is the index of modulation. This index determines the amount of frequency deviation (plus and minus) from the carrier frequency. In FM tape systems, X and f_c are standardized with f_c changing with record speed and $X = 0.40$. With $X = 0.40$ we see that a sine wave of full scale amplitude gives change in the carrier frequency from $0.6f_c$ to $1.4f_c$. It can be shown that it requires a bandwidth, B, of four times the frequency of interest for the FM process to reproduce it faithfully. Thus the highest usable frequency with the FM process at a given record speed is $B/4$ which must be less than $0.6f_c$ so as not to interfere with the lowest carrier frequency. Finally, from (5-5) we see that when $f_1 = 0$

$$f(t) = A_c \cos (2\pi f_c t + X) \tag{5-6}$$

Hence FM can be used to record a dc voltage.

FM recording electronics are supplied by the recorder manufacturer as plug-in units which can be interchanged with the direct record/reproduce units. The FM record electronics contain an amplifier, an oscillator whose frequency varies with the data signal, and an output amplifier to provide an impedance match for the record head. The signal is recorded to saturation on the tape at a constant amplitude and no bias is used since the information is obtained from a change in frequency, not amplitude. Each tape speed requires a different center frequency, and in the record amplifier this is usually provided for with plug-in or electrically switchable frequency determining units.

The FM reproduce electronics contain an amplifier, a limiter, a FM demodulator, a low-pass filter, and an output amplifier to increase the signal to a standard output voltage level and impedance. The first amplifier is not critical since the saturated signal already has an excellent signal-to-noise ratio on the tape. A limiting (clipping) operation is used to eliminate amplitude variations and provide a standardized signal to the demodulator. The demodulator has an output voltage that varies linearly with frequency. As with the FM record electronics, plug-in or electrically switchable elements are used to set center frequencies and the low-pass filter characteristics.

It is important to note that FM makes stringent demands on the ability of the tape transport to move tape across the heads at a precisely uniform rate. Any speed variations introduced into the tape at its point of contact with the heads will cause an unwanted modulation of the carrier frequency and result in noise.

5-5.5 Tape-Recorder Heads

The basic construction of both record and reproduce heads is similar, each consisting of a magnetic core on which is wound a number of turns of wire. The core consists of two C-shaped half sections which are made from a number of bonded laminations of ferromagnetic material. The surfaces of the half sections which interface with each other are lapped and polished, and the gap material is deposited on one. The two cores and their windings are then joined to form a head.

One head is used for each track of the tape. In a multitrack operation, several heads are assembled with intertrack shields to form a head stack. Extreme precision is used to align the heads in a head stack so that phase information (time-displacement error) can be obtained between channels. Precision mounting of the heads is necessary to reduce the short wavelength losses due to azimuth misalignment. These losses occur when the reproduce-head gap is not precisely parallel to the record-head gap and are defined by

$$\text{loss} = 20 \log_{10} \left(\frac{\sin x}{x} \right) \text{ dB} \tag{5-7}$$

where $x = (\pi W/\lambda) \tan A$, W is the track width, A is the angle of misalignment, and λ is the wavelength. Standardization in track and head geometry guarantee compatibility between instrumentation tape recorders; that is, tapes recorded on one manufacturer's machine will play back on another. From (5-7) it is found that for $W = 0.05$ in. the loss will be less than 0.1 dB for misalignments up to $1°$ if $\lambda \geq 0.01$ in. For misalignment less than $10'$ ($\frac{1}{6}°$) the loss will be less than 0.3 dB for $\lambda \geq 0.001$. This loss further restricts the high-frequency response of tape recorders.

5-5.6 Cross Talk

Cross talk is an undesirable characteristic of a multitrack recording system in which an output appears on one track as a result of the recording on another. Cross talk may result from simple coupling between individual track structures in the reproduce head and may be reduced with proper shielding and design. Similar coupling in the recording head can produce a record field at the gap of an adjacent head. This field alone is usually not strong enough to provide a recording. However, it can combine with any other field at the gap to produce a degree of recorded cross talk.

Cross talk also results from the lateral sensitivity of the reproduce head. The attenuation from distance between heads is given by $K_c \exp\left[-2\pi d/\lambda\right]$, where K_c is a constant related to the head shape and is of the order of unity, d is the distance from the edge of one track to the nearest edge of another, and λ is the recorded wavelength. Clearly, this type of cross talk becomes most severe at long wavelengths (low frequency) and thus adds to the general problem of directly recording long wavelengths without modulating a carrier.

5-5.7 Tape-Transport Mechanisms

The purpose of a tape transport is to move the tape past the heads at a constant speed and to provide the various modes of operation required for tape handling, without straining, distorting, or unduly wearing the tape. To accomplish this, a transport must guide the tape past the heads with precision and maintain the proper tension within the head area to obtain adequate tape to head contact. Transport mechanisms vary with manufacturer, but basically all systems consist of various guides to guide the tape around or across the record and reproduce heads. A capstan controls the speed of the tape across the heads. A pinch roller is used to maintain tension on the tape before and after contact with the capstan. These methods include variable speed motors and spring systems.

Since the capstan is used to control the tape speed it must run at a constant speed. One method is to use a hysteresis synchronous motor. Such a motor runs phase-locked to the ac power-line frequency and its long-term speed stability is thus as good as the power-line frequency stability (if operating with a constant load). Depending on power-line frequency for speed stability in many applications is a strong disadvantage. Even in laboratory use some instability can be expected. A metropolitan power source may show extremely good long-term stability but short-term variations as high as $\pm 0.25\%$ are not uncommon.

There are many methods to avoid use of power-line frequency for stability. However, if precise reproduction of recorded data is required a servo speed control is necessary in the reproduce mode. There are several techniques used for this and they all operate from a reference signal recorded on the tape with the data signals. For example, if the recording machine is driven by a hysteresis synchronous capstan motor, its precision 60-Hz supply is used as the reference signal and is recorded by modulating a 17,000-Hz carrier (per IRIG). If a dc capstan motor powers the recording machine the 60-Hz reference is generated from the speed determining reference oscillator and recorded in the same manner. Thus either system provides a reference signal which can be used by either system. Servo control of a hysteresis synchronous reproduce machine is accomplished by recovering the reproduced 60-Hz reference signal and comparing it with the local precision 60-Hz reference voltage. The putput of the comparator is a dc or very low frequency ac signal which controls the output of a 60-Hz voltage-controlled oscillator. This output is then amplified and used to drive the capstan motor. The servo action effectively locks the recorded reference signal to the local precision reference signal.

Short-term speed variations which are uniform across the tape can be caused in many ways in a tape-transport mechanism. Some of these are pulsations of the torque motors, reel eccentricities, and vibrations in the tape. These velocity variations have been variously described as wow, flutter, and drift. Flutter denotes variations in speed which occur at frequencies above 10 Hz; wow includes those between 0.1 and 10 Hz; and drift is used for those frequencies below 0.1 Hz. As applied to instrumentation recorders, common usage has broadened the definition to the term "flutter" to include all variations in the frequency range 0.1 Hz to 10 kHz. In general, the flutter spectrum is in the form of white noise. The percentage of flutter when expressed as the maximum peak variation of the signal frequency in terms of the average frequency is given by

$$F = \frac{f'' - f'}{f_a} \times 100\%$$

where F is the percentage flutter, f'' is the maximum frequency, f' is the minimum frequency, and f_a is the average frequency. For example, if a 7.5-kHz tone is recorded at $7\frac{1}{2}$ ips, the recorded wavelength will be 0.001 in. Suppose the deviation is $\pm 1\%$ (0.075 ips positive and negative), then the speed will change from 7.575 to 7.425 ips. The signal frequency will vary from 7575 to 7425 kHz.

The most widely used flutter measurement technique makes use of the conventional electronics in the FM recording units. The FM record amplifier, with its input shorted, is used to record a reference frequency and the FM reproduce unit is used to recover the average and instantaneous frequency variations of the reproduced signals.

Since flutter in an FM system constitutes a noise modulation of the carrier, it can be considered a measure of the dynamic range of the system. If flutter measurements are to be compared, one should make certain that they are measured in the same way; that is, in the same bandwidth and with the same type of detection: rms or peak-to-peak. Most manufacturers use the IRIG standard for the measurement of flutter.

5-5.8 Digital Recording

The digital recording process has become important as a result of the widespread application of digital computers to electronic data processing

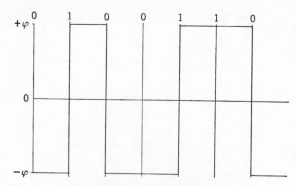

Figure 5-10 NRZ digital recording.

systems. As is discussed in detail in Chapter 7, an analog signal is sampled and the samples are then converted into a binary code. A binary code consists of a series or group of binary digits employing only two digits denoted 0 and 1. The actual voltage levels of the digits is not important since all that is required on the part of the electronic circuitry is a simple yes-no decision. Thus digital recording is accomplished by magnetizing the tape to saturation in either of its two possible directions (+ or −) at discrete points along its length.

Any of several basic techniques and variants may be used for recording binary digits. One of the most common is non-return-to-zero (NRZ). In NRZ recording the tape is always saturated in one direction or the other, the reversal in polarity occurring each time a change in bits occurs. For example the number 0100110 in NRZ notation is shown in Figure 5-10. The binary digits are usually recorded in parallel; that is, in a five-bit system all five bits are recorded simultaneously on five channels. Two other channels are required for timing and parity checking, as will be explained subsequently. The major advantage of the digital recording process is that it places no arbitrary limit on the accuracy of the system. The dynamic range of the tape, its linearity, or the type of signal-to-noise problems encountered in the other recording processes are not important. It is only necessary to properly encode the measurements in digital form and to reliably record and play back the corresponding pulses. Thus speed stability of the transport is not important since relatively large amounts of flutter can be tolerated without affecting the recording accuracy.

Since all the information is contained in the presence or absence of pulses on playback, the loss of pulses or the generation of spurious pulses caused by the tape imperfections cannot be tolerated. For this reason, special precautions are taken in the manufacture, inspection, and selection of tape intended for digital recording. Thus in general instrumentation tape is not used for digital recording. This does not solve completely the problem. As discussed under direct recording, tape dropouts become most critical at short wavelengths which approach the size of the gap in the reproduce head. This indicates a practical minimum duration for pulses, the number of pulses per inch of tape that can be recorded.

There are safeguards which are often built into the digital-recording process to provide greater reliability against the possibility of tape dropouts or other errors. One of these is the use of redundancy, in which the same information is recorded twice. A second method is the use of a parity check, in which one track on the tape is reserved for a pulse which is derived from the pulses being recorded simultaneously on the other tracks. A parity pulse is recorded of such a polarity that the sum of all bits on playback (including the parity bit) will be an odd number. Thus an error will be indicated if one (or an odd number) of the pulses are lost. It will not detect two simultaneous errors or an even number of errors.

A second problem is that of tape skew. This is any tendency for the center line of the tape to depart from perpendicularity with the record and reproduce head gaps as shown in Figure 5-11. In an exaggerated form the effect of tape skew would have the top and bottom head tracks reading bits from different channels corresponding to different words (combination of n parallel bits) at any given instant, instead of reading the bits corresponding to a single word. Skew is perhaps the most important single factor in pulse-packing density since the closer the pulses are packed together, the greater the possibility of errors due to tape skew.

5-5.9 Comparison of Recording Techniques

It was shown that the output from the direct-record process was directly proportional to frequency, azimuth angle, and the lifting of the tape from tape head. The direct process is also sensitive to tape coating thickness and surface roughness. When a wavelength is large enough so the full thickness

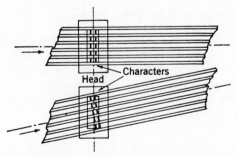

Figure 5-11   Error caused by tape skew.

of the coating is effective in producing output, any variations in thickness causes changes in the maximum available flux and thus in the output. Also anything which changes the separation between the head and the tape causes a loss in output which increases rapidly with decreasing wavelength. Dust, dirt, and oxide scrapings often come between the head and tape, causing a temporary reduction in the output level.

The result of the combination of all of these factors is that the absolute amplitudes as reproduced by a direct-record system may vary as much as $\pm10\%$, with the greatest variation occurring at the shortest wavelength. However, the bandwidth which can be recorded is much larger than is possible with FM recording. Therefore, direct recording is used where maximum bandwidth is required and amplitude variation errors are not critical. Generally, direct recording is used only for timing and for recording high frequencies which cannot be handled using other recording techniques.

FM recording overcomes some of the basic limitations of the direct process at the cost of a reduction in the high-frequency response. FM recording achieves dc response, improves amplitude accuracy, and generally gives a more uniform frequency response. The FM system is limited in bandwidth when compared to direct recording, dependent on constant tape speed, and has more complex electronics. Therefore, the FM recording process is used when the dc component of the input signal is to be preserved, or when the amplitude variation of the direct-recording process cannot be tolerated.

Digital recording is accomplished by magnetizing the tape to saturation in either of its two possible directions at discrete points along its length. Since timing information is included in the recording, the digital-recording process is insensitive to flutter. Dynamic range and linearity depends upon the number of digital bits that represent the input signal and are generally independent of the tape-transport system. Since all information is contained in the presence or absence of pulses, the systems sensitivity to tape dropout errors (lack of oxide coating) is most critical.

5-5.10   Manipulation of Data Using Magnetic Tape

Magnetic tape is useful in that the signal information is preserved in its electrical form so that the original event can be recreated at any future time. Additionally, events can be recreated on playback either faster or slower than they actually occurred, with a resulting increase or decrease of the signal frequency spectrum. These two properties are very useful in processing signals.

As was discussed in Section 1-4.5 a spectrum analysis of an aperiodic signal could be performed by "capturing" the aperiodic signal on tape and then repeating it in its entirety at some periodic interval. Hence the aperiodic signal is recorded and the section of tape which contains the signal is found during playback. This section is then cut from the reel and spliced to form a tape loop. Most tape-recorders manufacturer sell tape loop adapters which are mechanisms for handling a tape loop. In general, the loop adapters replace the reels used for normal reel-to-reel recording. The mechanics of splicing the tape requires care, since a poor splice will generate an additional signal when it passes by the tape head. To overcome this problem some tape recorders have tape splice blankers which squelch the output from the splice. The squelch is initiated from a signal on another channel of the tape.

To be able to properly analyze a periodic signal (aperiodic made periodic) it is necessary that the distance between the spectral lines be larger than the analyzing bandwidth (recall Figure 1-7). If the duration of one period of the repetition frequency is T_0 and the duration of the pulse is $2t_0$, then the ratio $2t_0/T_0$ is subject to strict limitations. As can be seen from Figure 1-7 the zero in the theoretical pulse spectrum occur with frequency intervals $1/2t_0$. To be able to obtain more than one spectral line between successive minima, the ratio $2t_0/T_0$ must therefore be less than unity. Experiments have shown that five lines between minima seem to give the spectrum. However, the smaller the ratio $2t_0/T_0$ becomes the smaller the available dynamic range for the analysis becomes. A ratio too-small must be avoided because of crest-factor limitations in the analyzing equipment. As a practical compromise, a $2t_0/T_0$ ratio between 0.2 and 0.33 is usually used.

Since a signal or an event can be recorded at one tape speed and played back at a different speed, signals can be analyzed with instruments that do not have the frequency range of the signal being recorded. This is accomplished by increasing the frequency of very-low frequency signals by recording at a slow speed and playback at a higher speed and the opposite for a high-frequency signal. The frequency transformation is directly proportional to the change in tape speeds.

An additional use of tape-speed transformation is the decrease in analysis time. Recall (1-150) where it was shown that the normalized standard error associated with a spectral density measurement was given by $\epsilon = 1/\sqrt{B_e T}$. If the tape speed during reproduce is increased by a factor k, then for the same resolution $R = kB_e$ that would be obtained without tape speedup, the equivalent averaging time is $T = 1/R\epsilon^2$. Thus the time required to perform the analysis for the same normalized standard error and resolution has been reduced by the change in tape speed k. If the same analysis time and resolution were used, the normalized standard error would be reduced by $1/\sqrt{k}$.

Application 5-1   Measurement of Aircraft Fly-Over Noise

The judgment of loudness varies nonlinearily with amplitude and frequency. Various weights have been developed to relate measured spectrum to subjectively judged annoyance. The most commonly used weight for

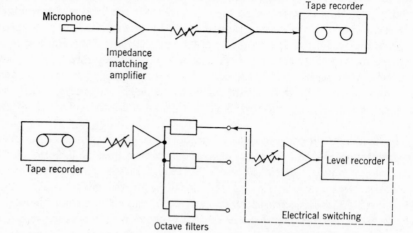

Figure 5-12 Data collection and analysis system for the measurement of aircraft flyover noise.

intermittent noise is the PNdB. The PNdB requires an octave- or $\frac{1}{3}$-octave analysis in the frequency range of 50 to 12,500 Hz. For the measurement of annoyance due to aircraft flyover noise, data are taken at a number of locations simultaneously and the PNdB is computed. The most common method is to use portable data-gathering equipment and perform the actual analysis back in the laboratory.

The block diagram of the data-collection and analysis system is shown in Figure 5-12. The data-collection system consists of a condenser microphone, cathode follower, amplifier with stepped attenuators, and tape recorder. A detector and voltmeter is also included to measure the output voltage from the amplifier into the tape recorder. In order to calibrate the entire system a portable acoustic sound source is used. The sound source, called a piston phone, gives a sinusoidal acoustic output of known frequency and level. The entire system is calibrated at the beginning and end of each tape. This calibration is recorded and establishes a reference level for later analysis. When an aircraft flys over a recording is made and the gain of the amplifier is noted. Since the gain of the amplifier was known during calibration together with the calibration level it is an easy matter to establish the level of the signal during analysis.

To analyze the data either the $\frac{1}{3}$-octave real-time analyzer discussed in Section 7-16.2 can be used or a tape loop is formed and the analysis system shown in Figure 5-12 is used. The first amplifier is used to increase the magnitude of the signal from the tape recorder. The amplified signal is then filtered with the contiguous set of octave- (or $\frac{1}{3}$-octave) filters. The second amplifier is used for impedance matching. The level recorder is used to record the data and advance the filters by closure of a microswitch. Since the noise level is measured to determine annoyance; the detector time constant depends upon the response time of the ear and is not chosen for statistical accuracy. It has been found that the response time of the ear is approximately 50–200 msec. Thus for $\frac{1}{3}$-octave analysis a 60-msec (or somewhat greater) averaging time is used. The data are recorded and the maximum rms value is used for PNdB calculations. The final results are noise contours of the flyover noise at various locations.

Application 5-2 Structure of Low-Speed Turbulence

The measurement of the structure or properties of turbulence was already discussed in Application 1-2. However, the correlation delays in low-velocity turbulence are extremely long, therefore requiring lengthy processing times. In order to reduce this time the output of the anemometers are recorded at $1\frac{7}{8}$ ips on an FM tape recorder. The FM recorder is used because it will record very low frequency signals. The recorded signal is then played back at a faster speed, say 15 ips. In this case, for the same statistical accuracy at $1\frac{7}{8}$ the analysis time has been reduced by a factor of 8.

The same technique could be used if it were desired to measure the frequency spectrum of the turbulence. Most frequency analyzers have a lower limit in the audio range. If, for example, the lower limit of the filter was 6.3 Hz, a signal of 0.8 Hz could be analyzed by recording at $1\frac{7}{8}$ ips and playing back at 15 ips. Proportionately lower frequencies could be considered if higher playback speeds are available.

5-6 Oscilloscope

5-6.1 General Description

A simplified block diagram of a cathode-ray oscilloscope is shown in Figure 5-13. A narrow beam of electrons is projected from an electron gun through a set of horizontal and vertical deflection plates. Voltages applied to these plates create an electric field which deflects the electron beam and causes horizontal and vertical displacement of its point of impingement on the phosphorescent screen. The phosphorescent screen emits light which is visible to the eye.

The most common mode of operation is when the oscilloscope is used to view voltage variations with time. In this mode the horizontal plates are driven with a ramp voltage which causes the spot to sweep from left to right at a constant speed. A voltage amplifier is simultaneously used to drive the vertical plates causing the spot to move up and down, thus giving a trace.

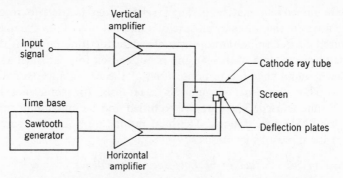

Figure 5-13 Simplified functional diagram of the cathode-ray oscilloscope.

It is also possible to use a horizontal voltage amplifier to obtain an *X-Y* recording.

Some oscilloscopes may have two vertical beams and two amplifiers (termed dual beams) or the vertical preamp may be a dual trace. The dual-trace preamp consists of two amplifiers and a switching network as shown in Figure 5-14. The switching circuit (termed a chopper) consists of a switching multivibrator and switching circuit. In the chopped mode channels *A* and *B* are sampled at a high rate (e.g., 500 kHz). Since the preamp is limited in the chopped mode to frequencies much less than 500 kHz an alternate mode is also included for higher frequencies. In the alternate mode one sweep is made with channel *A* and then one with channel *B*. The alternate rate depends upon the sweep rate of the oscilloscope. Thus the alternate mode is usually used at high frequencies while the chopped mode is used at low frequencies.

5-6.2 Schmitt Trigger

Consider the simplified block diagram of the time base. Note that there is no provision for initiating the sawtooth signal with respect to an input signal. For example, consider the sine wave and the sawtooth shown in Figure 5-15. Notice that the signal will appear on the oscilloscope at a different time

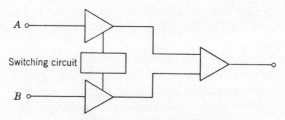

Figure 5-14 Dual-trace amplifier.

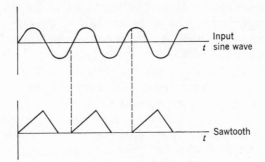

Figure 5-15   A sinusoidal input and the time-base sawtooth.

during each sweep. Thus the signal will appear to move across the screen. It is apparent that the sweep should be initiated by the input signal.

One circuit that may be used to initiate the sawtooth time base is called a Schmitt trigger (recall Section 3-11.6). The Schmitt trigger has two stable states, V_1 (termed "on") and V_2 (termed "off"). Assume that the input to the Schmitt trigger is some voltage greater than E_2. In this case the trigger is at the "on" state and the output of the trigger is V_1. If the input voltage is now lowered to E_1 (where $E_2 > E_1$) the output of the trigger is V_2. The difference between E_2 and E_1 is termed the hysteresis range of the trigger. A Schmitt trigger is termed a bistable circuit and its action is shown in Figure 5-16.

To change the Schmitt trigger from the first stable state (output V_1) to the second stable state the input voltage must be E_1 or less. Thus to switch the trigger from one state to the other and back again, a periodic wave form whose amplitude varies from greater than E_2 to less than E_1 must be applied. This actuating signal (including the dc level of the actuating signal) must at times rise above the upper hysteresis limit E_2, and at other times fall below the lower hysteresis limit E_1. The output signal may or may not be symmetrical, depending upon the wave form and the dc level of the actuating signal. It is apparent that the actuation level of the trigger can be

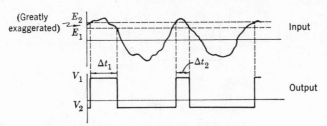

Figure 5-16   Input to and output from a Schmitt trigger.

changed by adjusting the dc level to the input of the trigger. Also, the actuation level of the trigger may be adjusted by changing the dc voltage levels within the trigger itself. A very important application of the Schmitt trigger is its use as an amplitude comparator to mark the moment at which an arbitrary wave form attains a particular reference level. This circuit has the advantage that when the comparison point is reached, the comparator output makes an abrupt and pronounced change.

For some purposes it is necessary to make the trigger operate in a free-running manner. That is, it is required that the circuit deliver a recurrent output wave form even when no input actuating signal is applied to the input. When this happens the oscilloscope is said to be in the AUTO mode. In this mode the sweep will trigger after one sweep plus some delay, of the order of 50 msec. However, when a suitable input signal is applied to the trigger input it will override the auto circuitry and trigger the sweep.

Since it is necessary to start the sweep only when the signal is going positive (positive slope) or negative (negative slope), a differentiator and diodes are used to block either positive or negative pulses following the trigger circuit. This circuit is called the trigger shaper. The output of the trigger shaper initiates the sawtooth by means of a sweep gate. The sweep gate also allows the sawtooth one complete sweep before another pulse can initiate the sweep. The circuit that senses that the sawtooth is sweeping is called a "hold-off circuit." It is also possible to switch the "hold-off circuit" to single sweep so that only one trace is obtained after the sweep has been initiated. The complete block diagram of the time-base generator is shown in Figure 5-17.

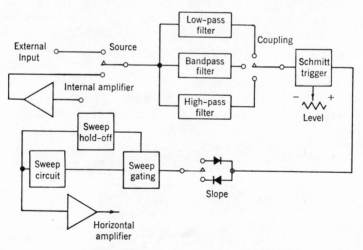

Figure 5-17   Oscilloscope time base.

Note that there are four controls on the block diagram; the SOURCE, SLOPE, LEVEL, and COUPLING. The SOURCE determines the origin of the signal used for triggering. The signal for triggering may be selected from the vertical amplifier (internal), be connected into the external trigger input separately (external), or be a signal taken directly from the oscilloscope transformer, normally 60 Hz (line frequency). The 60-Hz source is very useful for checking for ground loops since the signal will appear stationary if the signal is related to the line frequency. The SLOPE control determines whether a trigger signal will be generated on a positive-going or negative-going signal. The LEVEL control determines the level at which the trigger circuitry will generate a trigger signal. Note that if the LEVEL is set for E_2 volts and the trigger SLOPE is set to positive, a signal going from 0 to $E_2(E_2 > E_1)$ will trigger the sweep, but a signal going from E_2 to E_1, even though the signal passes through the desired level, will not because of the trigger shaper. Trigger COUPLING determines the way in which the signal is connected to the trigger circuit. The signal may be coupled directly to the trigger (dc). This is necessary for triggering on phenomena that vary slowly. The signal may be coupled through a bandpass or high-pass filter. The advantage of the other couplings is that dc levels present in the trigger signal are blocked.

Notice that if a dual-trace preamplifier is used and phase information between channels is to be preserved either an external trigger must be used or the preamplifier must be in the chopped mode. If an internal trigger is used and the preamplifier is in the alternate mode each signal would establish the same level for triggering even though they might be out of phase. Thus the signals would appear in phase when viewed on the oscilloscope. Some oscilloscopes have provisions for taking the trigger signal from only one input. In these oscilloscopes phase may be measured in the alternate mode with a trigger from the internal amplifier.

5-6.3 Delayed Sweep

It is often necessary to examine only a small portion of a signal. This may be required for the determination of rise time or phase-angle measurements. It may be difficult to trigger the oscilloscope such that this portion of the wave form can be examined. Further, if the amplitude of the time base generator is increased (usually termed magnification), any nonlinearities of the time base are, in effect, increased by the amount of magnification. Thus delayed sweeps are used.

Delayed-sweep measurements are based on the use of two linear calibrated sweeps (time bases). The first sweep, termed the delaying sweep, allows a specific delay time. When this time is reached, the delayed sweep, from a second time base, starts. If the delayed sweep is three decades faster than

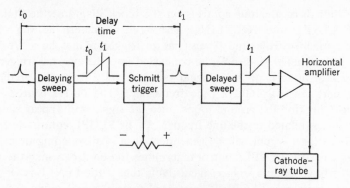

Figure 5-18 Delayed sweep generator.

the delaying sweep a $1000\times$ magnification has been achieved without increasing the nonlinearities of the time base 1000 times.

The block diagram of a delaying-sweep oscilloscope is shown in Figure 5-18. When a trigger pulse is applied to the delaying sweep at time t_0, the delaying sweep is started. The delaying-sweep ramp is applied to a Schmitt trigger which produces a trigger pulse at a later point in time t_1. Note that this time occurs when the ramp is equal to a preset Schmitt trigger voltage. The comparator is set by a ten turn potentiometer calibrated in divisions of 0.01 of full scale. The amount of delay is then the delay of the time base multiplied by the value of the potentiometer. The trigger pulse occurring at t_1 starts the delayed sweep. Delay time is defined as the difference in time between the start of the delaying sweep and the start of the delayed sweep and can be expressed as $t_1 - t_0$. Note that the delay depends upon the sweep rate of the delaying sweep (usually termed time base B) and the voltage set on the voltage comparator (usually termed delay-time multiplier). When a delayed sweep receives a pulse the second time base generates a ramp function for sweeping the oscilloscope trace.

Since large magnifications are possible with the delayed-sweep generator it is convenient to determine the amount of delay and the delayed sweep period with respect to the signal displayed on the oscilloscope rather than through calculation. For this reason an intensification mode is included in the delayed-sweep oscilloscope. An intensifying pulse indicates where the delayed sweep starts with respect to the delaying sweep, and so delay time can be determined independently of horizontal amplifier and cathode-ray tube considerations. The intensifying pulse is applied to the electron gun to increase the intensity of the delayed sweep portion of the signal. The block diagram of the delayed-sweep oscilloscope in the intensification mode is shown in Figure 5-19.

The wave forms shown in Figure 5-20 represent the wave form that will be magnified in the delayed-sweep mode. The oscilloscope has been set to the

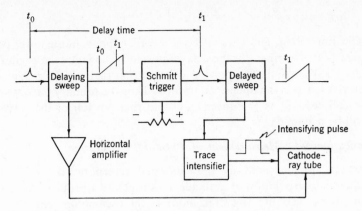

Figure 5-19 Delayed-sweep oscilloscope in intensification mode.

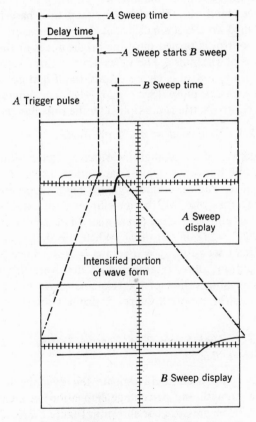

Figure 5-20 Wave form as seen during intensification mode and delayed sweep mode.

intensification mode. The actual delayed sweep is shown in the lower portion of Figure 5-20. The time t_0 is the delay time and is shown by the beginning of the intensified portion. The portion of the signal that will be magnified by the delayed sweep is the intensified portion. Thus the primary purpose of the intensification mode is to conveniently find that portion of the wave form that will be magnified.

Application 5-3   Measurement of Pulsewidth (Duration)

If pulses are measured by the calibrated screen on the oscilloscope, parallax and interpolation of graticule lines becomes a problem for accurate measurements. The intensification mode of the oscilloscope can be used to eliminate these errors. The oscilloscope time base and trigger are first adjusted to obtain at least one complete cycle of the wave form. The intensification mode is then used and the intensity adjusted to make the intensified part of the trace stand out. The delayed sweep time base is then adjusted until the intensified area is a small bright dot. To measure pulse duration, the delay time control is turned until the bright dot is at the 50% point on the leading edge of the pulse. The delay time control setting is recorded. Then the delay time control is changed until the bright dot is moved to the trailing edge of the same pulse. (See Figure 6-12). The difference between the two delay time control settings multiplied by the time base is the pulsewidth.

Application 5-4   Measurement of Phase Angle

A common method of phase-angle measurement cross-plots two sinusoidal signals of the same frequency against each other, using an X-Y plotter for very low frequencies and an oscilloscope for high frequencies. As shown in Figure 5-21 the cross plot will be an ellipse, and suitable measurements obtained from this ellipse give the phase angle. Let the two inputs be $s_i(t) = A_1 \sin \omega t$ and $s_2(t) = A_2 \sin(\omega t + \varphi)$. When $t = 0$, $s_1(0) = 0$ and $s_2(0) = A_2 \sin \varphi$. Then if we let $s_2(0)|_{s_1(0)} = B$, $\sin \varphi = B/A_2$. Since B has two values $(+, -)$ the quadrant of φ is ambiguous; however, this can usually be resolved by visual observation of the two sine waves plotted against time, that is, on a dual-beam oscilloscope or from knowledge of the system characteristics.

Application 5-5   High-Strain-Rate Tensile Testing of Materials—Split Hopkinson Pressure Bar

An apparatus often used to investigate the dynamic tensile properties (ultimate tensile strength and percentage elongation) of a class of materials is the split Hopkinson pressure bar shown in Figure 5-22a. The compressed air gun is used to propel a cylindrical projectile through the barrel. Two photosensitive devices are used to measure the projectile's velocity by having

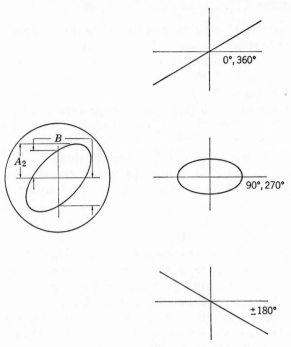

0°, 360°

90°, 270°

±180°

Figure 5-21 Phase angle from a Lissajous figure.

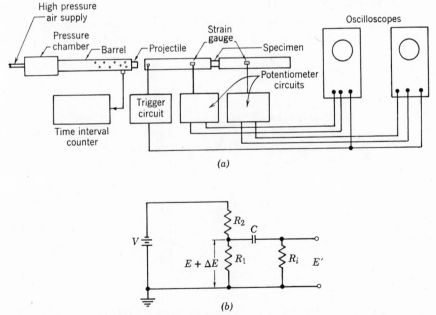

Figure 5-22 (*a*) Functional diagram of split Hopkinson pressure bar and (*b*) the potentiometer circuit.

the projectile break the optical path of the first device to trigger the counter and the second to stop the counter. The stress, strain, and strain rate in the specimen are a function of the incident and reflected strain in bar 1 and the transmitted strain in bar 2. Hence strain gauges are positioned at a point on bar 1 and at a point on bar 2. The read-out devices are two oscilloscopes. The oscilloscope sweeps, which are in the single-sweep mode, are initiated by a trigger pulse from either another strain gauge or photosensitive device positioned at the beginning of bar 1. The oscilloscope traces are photographed and from these pictures the pertinent data are obtained.

The output of the strain gauges are fed into a simple potentiometer circuit shown in Figure 5-22b. It can be shown that the time-varying output of this circuit is

$$E' = \frac{\omega \tau_i V}{\sqrt{1 + \omega^2 \tau_i^2}} \frac{r \epsilon'}{(1 + r)^2} GF$$

where $\tau_i = R_i C$, $r = R_2/R_1$, GF is the gauge factor [recall (3-56)], and ϵ' is the strain. From (2-51) we see that E' as a function of ω behaves as a high-pass filter. From Figure 2-12b one can determine the lower limiting frequency for negligible attenuation due to $R_i C$. To determine the loading effect of R_i on the circuit we use (3-13b) or Figure 3-11 and multiply the percentage error obtained from either of these by $2r/(1 + r)$ assuming that $R_1/R_i \ll 1$. A typical value of r is of the order of 9.

5-7 Fiber-Optic Oscilloscope

The fiber-optic oscilloscope is basically a single-channel oscilloscope with fiber optics to enable recording on direct-print paper. Direct-print paper

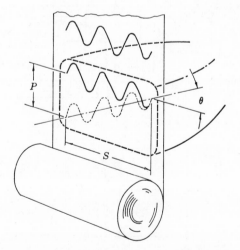

Figure 5-23 Transverse recording of a fiber optic oscilloscope with skew correction.

does not require developing to produce an image. The light output from the cathode ray-tube is focused on a continuous strip of direct-print paper by the fiber-optics system. The paper may be moved contiguously or continuously. In the continuous paper mode a skew correction is made electronically to correct for the motion of the paper. The results of the correction and the uncorrected signal are shown in Figure 5-23. All normal controls found in oscilloscopes, such as trigger level, slope, and coupling, are also used. The advantage of the system compared to an oscilloscope and camera is the fact that continuous recordings can be obtained.

The bandwidth of the system is typically dc to 1 MHz with sweep rates of 1 μsec/cm to 1 sec/cm and paper speeds of 0.1 cm/sec to 250 cm/sec. The X-Y recording may also be obtained by using a built-in horizontal amplifier rather than the time-base amplifier.

5-8 Summary

Recorders are used to obtain a permanent record of an event. The record may be obtained as a function of time or, with some recorders, as a function of another variable. All ink-type recorders have a mechanical system that is required to move and hence are limited to low frequencies, unless detectors are used. A recorder that contains a detector and a logarithmic converter is generally called a graphic level recorder and may be used with other equipment to obtain the spectrum analysis or the transfer function of a system automatically. Since a detector is used the upper limiting frequency of the level recorder may be of the order of 200 kHz.

Magnetic tape recorders are used to store data. The signal information is preserved in its electrical form so that the original event can be recreated at any future time. Additionally, tape provides the ability to alter the time base. This permits events to be recreated on playback either faster or slower than they actually occurred, with resulting multiplication or division of the signal spectrum. A tape loop may be used with a tape recorder for analyzing aperiodic data. There are three major recording methods: direct, FM, and digital. Direct recording has higher frequency response than any other type; however, it will not record a dc signal and is subject to data loss due to tape drop-outs. FM recording will respond to dc, is not affected by tape drop-outs, and has a more uniform frequency response. The FM system is extremely sensitive to tape-transport-speed regulation and requires complex electronics as compared to a direct-record system. Digital recording will respond to dc and is insensitive to tape-transport-speed instabilities. The accuracy of the system depends on the number of bits recorded. Digital recording has poor tape economy compared to FM and its reliability is extremely dependent on tape quality.

Oscilloscopes have the greatest bandwidth of any recording method. In general, they are used to measure amplitude of a signal as a function of time, although they may be used to measure amplitude as a function of a variable other than time. Most oscilloscopes include a time base which must be initiated by some stable property of the wave form. Thus there is the provision for changing trigger levels, slopes, and sources. Some time bases have a delayed sweep capability. The delayed sweep is very useful when measuring properties of pulses, since large magnifications can be obtained without magnification of any inherent nonlinearities of the time base.

References

1. J. T. Broch, "R.M.S. Recording of Narrow Band Noise with the Level Recorder Type 2305," B & K Instruments, Cleveland, Ohio, Technical Review, No. 4, 1960.
2. G. L. Davis, *Magnetic Tape Instrumentation*, McGraw-Hill Book Company, Inc., New York, 1961.
3. E. C. Doebelin, *Measurement Systems: Application and Design*, McGraw-Hill Book Company, Inc., New York, 1966.
4. W. H. Evans, *Introduction to Electronics*, Prentice-Hall, Inc., Englewood Cliffs, N.J., 1962.
5. N. M. Hayes, *Elements of Magnetic Tape Recording*, Prentice-Hall, Inc., Englewood Cliffs, N.J., 1957.
6. J. A. Howard and L. N. Ferguson, "Magnetic Tape Recording Handbook," Application Note 89, Hewlett-Packard, Mountain View, Calif., 1967.
7. C. B. Pear, *Magnetic Recording in Science and Industry*, Reinhold Publishing Corp., 1967.
8. "IRIG Telemetry Standards," Document Number 106-66 Defense Documentation Center, Cameron Station, Alexandria, Va.
9. *Typical Oscilloscope Circuitry*, Tektronix, Inc., Beaverton, Ore., 1966.
10. "Service Scopes," Tektronix, Inc., Beaverton, Ore., No. 50, June 1968.
11. "A Convenient Method for Measuring Phase Shift," Application Note 29, Hewlett-Packard Co., Palo Alto, Calif.

6

SIGNAL GENERATORS

6-1 Introduction

In Chapter 1 it was shown that the transfer function of a system could be obtained with a sine wave, an impulse function, or white-noise input. Devices to generate these functions are termed signal generators. It is the purpose of this chapter to show how such devices are used and their limitations.

6-2 Sinusoidal Signal Generators

6-2.1 Feedback Oscillators

The purpose of the sinusoidal signal generator (usually termed oscillator) is to generate the function

$$e_{\text{out}} = A_0 \sin (2\pi f_0 t) \tag{6-1}$$

where A_0 is the amplitude of the signal and f_0 is the frequency in hertz. Since e_{out} is to be used for purposes of finding the response of some system to a sinusoidal input, the signal should not contain noise or harmonics of the fundamental frequency, f_0. Also, it is desirable that the frequency of the signal generator be known with reasonable accuracy without the aid of external equipment such as a frequency counter (see Section 7-10.1).

An oscillator may be considered as a form of a feedback amplifier in which special requirements are placed on the feedback loop represented by the complex quantity, β, and the amplifier gain, $A(\omega)$. Consider the circuit shown in Figure 6-1 where we have an amplifier that is operating in its linear amplifying modes. Its input will be supplied from its output, thus forming a closed-loop system. Assume that the system is operating as an oscillator and is producing a sinusoidal output voltage. A portion of this is fed back into the input. For equilibrium, the input must be just sufficient to produce the output. The gain around the loop is unity. A circuit that meets the above requirements is shown in Figure 6-2 and can be easily shown to perform double integration. However, note that the output of the double integrator has been connected to the input. The frequency of oscillations are given by

211

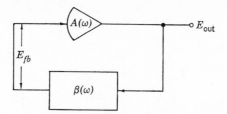

Figure 6-1   Basic feedback oscillator.

$f = 1/(2\pi RC)$. In general, the circuit is used with a buffering circuit to avoid loading the amplifier and to provide an adjustable output voltage.

In general, the frequency of this oscillator is continuously changed by changing the capacitance of the phase-shift circuit. Due to frequency limitations this allows a continuous frequency range of about 1 decade (e.g., 10 to 100 Hz). To obtain a larger frequency range a fixed component (in this case R) is changed which allows the oscillator to generate frequencies in other decades.

The oscillation frequency of an oscillator circuit may drift because of the variation of the active-device parameters, aging of the elements, variations in environmental conditions, biasing, and loading conditions. The active-device parameter variation may be reduced by negative feedback (see Section 3-8).

In general, frequency drift is inversely proportional to the Q of the tuned circuit. This is because the frequency deviation required in the loop transmission to yield 360° phase shift is inversely proportional to the Q of the loaded feedback circuit. In many oscillator circuits, where the frequency stability is an important factor, piezoelectric crystals are used in the tuned circuit. The equivalent circuit of the crystal is shown in Figure 6-3. The

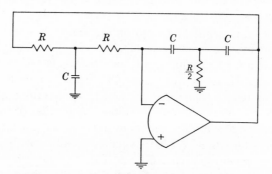

Figure 6-2   Double integrator circuit performing as an oscillator.

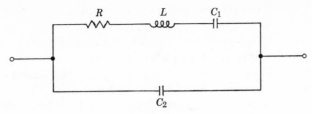

Figure 6-3   The equivalent circuit for a crystal used as a tuning element.

crystal has a very high Q; hence the frequency stability of crystal-controlled oscillators is excellent.

The variation of frequency with temperature is often expressed by the ratio $(\Delta\omega/\omega_0)/(\Delta T/T_0)$, where ω_0 and T_0 are the desired frequency of oscillation and the operating temperature, respectively. In some oscillators the requirement on the frequency drift due to changes in temperature is expressed as parts per million per degree centigrade.

6-2.2   Beat-Frequency Oscillator

To increase the continuous frequency range of an oscillator a device called a beat-frequency oscillator is used, which actually depends on the heterodyne principle rather than the beating principle. The beat-frequency oscillator has the added advantage that an external voltage may be used to control the output voltage of the oscillator. Consider the block diagram of the beat-frequency oscillator shown in Figure 6-4. One oscillator is fixed at a constant frequency ω_f with amplitude A_f. The output of the fixed oscillator

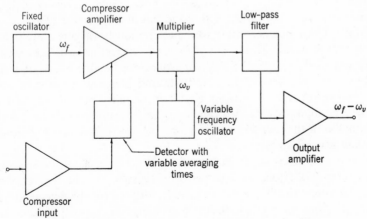

Figure 6-4   Functional diagram of a beat-frequency oscillator.

is passed through a compressor amplifier, as described in Section 3-11.10, and modified so that its gain can be changed from an external source. For purposes of explanation assume that the compressor amplifier has unity gain. The output of the compressor amplifier is connected to a multiplier. A second oscillator is also connected to the multiplier. The second oscillator is a variable-frequency oscillator and has a frequency range of ω_{vl} to ω_{vu} and an amplitude A_v. Usually ω_{vu}/ω_{vl} is less than ten; that is, less than 1 decade. As was shown in Section 2-4.4 and (2-62), the output of the multiplier is

$$f_m(t) = k_0 A_v A_f [\cos (\omega_f - \omega_v)t + \cos (\omega_f + \omega_v)t] \qquad (6\text{-}2)$$

The output of the multiplier is then passed through a low-pass filter with a cut-off frequency $\omega_c (\omega_v > \omega_c > \omega_f - \omega_v)$. The filtered signal is then amplified and connected to a stepped attenuator and variable potentiometer.

It is often desirable to vary the output of the oscillator as a function of frequency. Let us assume that one has at his disposal a transducer whose output is proportional to the physical quantity of interest. Further the output of this transducer is known (by prior calibration) to be constant over the frequency range of interest. The transducer is now mounted at the input to the system under investigation. Its output is then detected (recall Chapter 4), amplified, and fed to the compressor amplifier. Since the output of the compressor amplifier is inversely proportional to the detected output of the transducer, the input to the system under investigation can be maintained constant in the desired frequency range.

The relationship between the averaging times of the detector, T_0, used in the preceding and the frequency of the compressor input signal, f_{co}, should be given by $T_0 f_{co} > 1$. If this relationship is not obeyed the compressor circuit will work incorrectly or not at all. For example, when $T_0 \ll 1/f_{co}$ the compressor circuit will be correcting the output at the rate of f_{co}; in other words, it is correcting according to the continuously varying amplitude of a sine wave of frequency f_{co} instead of to its rms or rectified average amplitude. Conversely, if $T_0 \gg 1/f_{co}$ the detector would not be able to respond rapidly enough to the changes in the amplitude and hence would not be able to properly control the compressor circuit in certain applications. For these reasons a contiguously stepped time constant is available on beat-frequency oscillators with the compressor function. We see that at higher frequencies T_0 should become smaller.

To see that a decrease in T_0 is useful, we consider a moderately high Q system ($Q > 25$). If one were to perform the sweep of the frequency range of interest automatically, sufficient time is required for the detected input to the compressor amplifier to effectuate a change in the output level. In the neighborhood of the resonances the amplitude changes required by the compressor circuit are large (>20 dB). Therefore by judiciously selecting T_0 to be the minimum required for proper operation, great savings in testing time can be realized. This time constant, therefore, is often expressed in decibels per second: the amount of amplitude, in decibels, the unit can regulate per second.

Application 6-1 Sinusoidal Vibration Testing

Vibration testing is useful to determine a structure's response due to vibrations. In general, a sinusoidal-vibration-test system consists of an electrodynamic exciter (as described in Application 3-4), a sinusoidal signal generator with a power amplifier, and a transducer system to measure the input and output vibration level. The simplest sinusoidal vibration test uses one accelerometer mounted on the base of the exciter. In general, other accelerometers are mounted at various points on the object under investigation. The output voltage of the accelerometer is followed by a high-impedance unity gain amplifier (cathode or emitter follower), as described in Application 3-3 or Section 3-11.9, and then amplified and detected. The oscillator is set at one frequency, the output voltage is adjusted until the desired vibration level is achieved at the base of the exciter, and then the acceleration at the other accelerometers on the object are measured. A new frequency is set on the oscillator and the steps are repeated. The output level of the oscillator must be changed at each frequency because of the effects of resonant loading on the exciter and impedance mismatches between the power amplifier and the exciter. A typical acceleration level versus frequency measured on an exciter loaded by a single degree-of-freedom system is shown in Figure 6-5.

In order to decrease the time required to obtain a vibration response test a motor-driven beat-frequency oscillator may be used, as shown in Figure 6-6, to perform an automatic vibration test. As discussed previously, the correct detector averaging time (compressor speed) will depend upon the Q of the system. A level recorder is shown in the block diagram as a recording device although the combination of a detector, log converter, and X-Y plotter could be used as an alternate record system. If the system under test has a very high Q special precautions must be taken as described in Application 6-2.

It is sometimes necessary to perform a test using constant velocity or displacement rather than constant acceleration. All that is required is that the accelerometer voltage be integrated once to obtain the velocity or twice to obtain displacement before being connected to the compressor circuit. Note, however, that this can only be performed at relatively low frequencies since the amount of power required of the power amplifier and the amount of force required by the exciter increase directly with radian frequency for constant velocity and with the square of the radian frequency for constant displacement.

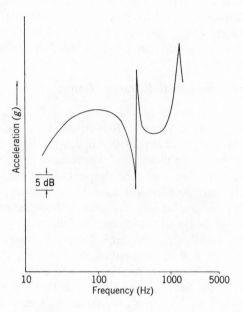

Figure 6-5 A typical acceleration level versus frequency for a vibration exciter with a single degree of freedom load.

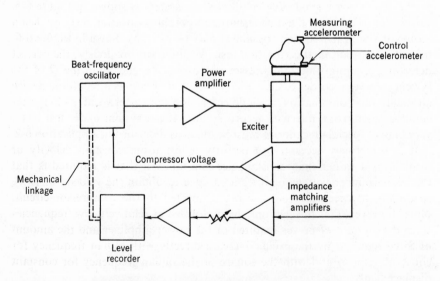

Figure 6-6 Automatic sinusoidal vibration test system.

6-2.3 Voltage to Frequency Converters (Voltage-Controlled Oscillators—VCO)

The conversion of a dc voltage input to a periodic wave output whose frequency is proportional to the dc input has several useful functions. Such devices, termed voltage-controlled oscillators, are widely used in FM telemetry systems, in integrating digital voltmeters where a dc signal is converted into a periodic wave of proportional frequency, in magnetic-tape recorders (see Section 5-5), and in sweep oscillators. In general, all voltage-controlled oscillators use some type of multivibrator. Before proceeding with a discussion of voltage-controlled oscillators it is helpful to consider multivibrators in general, even though they are not all directly applicable to voltage-controlled oscillators.

There is no universal distinction between multivibrators and trigger circuits, but for purposes of this discussion the classification suggested by Figure 6-7 will be used. Each of the circuits indicated has the same input signal, with the exception of the astable multivibrator which has none. The bistable circuit has two stable states and is shifted from one to the other alternately by each input pulse. The bistable circuit can exist indefinitely in either of the two states (see Section 7-6.2). Note that only half as many output pulses are generated as there are input pulses received by the circuit.

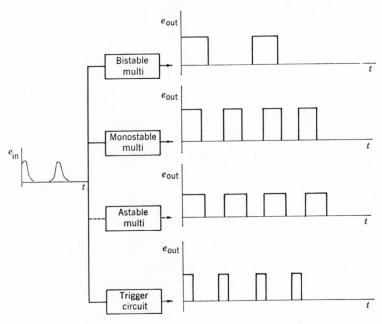

Figure 6-7 Trigger and multivibrator responses.

The monostable circuit has only one stable state and remains in that state for a fixed period of time independent of the input signal. The fixed period of time depends upon the components of the circuit. In particular, the quasi-stable time may be changed by varying the circuits passive elements (resistors and capacitors). The circuit requires a triggering signal to induce a transition from the stable state to the "on" state. The circuit may remain in its quasi-stable state for a time very long in comparison with the time of transition between states. Eventually, it will return from the quasi-stable state to its stable state, no external signal being required to induce this reverse transition. The astable circuit is, in reality, an oscillator and gives an output without an input signal. The circuit readily synchronizes with any input signal that will cause its period to decrease slightly. Thus the astable is forced to change states before it would naturally. Astables can be made to synchronize with input signals whose frequencies are lower by integers of one through ten and also frequencies that are greater by an equivalent range of integers. This circuit is often used for frequency multiplication or division. The switching frequency of the astable depends upon the passive elements in the circuit. Also, the frequency may be changed by applying an external voltage to the circuit. Thus the astable may be used directly as a voltage-controlled oscillator over a limited range. The trigger circuit is the same Schmitt trigger discussed in Section 5-6.2.

Voltage-controlled oscillators can be divided into two groups: oscillators for data transmission and oscillators for measurements. The former group include oscillators for telemetry and for FM tape recording. The latter group includes oscillators for integrating digital voltmeters and sweep oscillators. The reason for the division into two groups is the circuitry and and bandwidth requirements of each. The data transmission oscillator must deviate in frequency a small percentage about a center frequency. For the best resolution, oscillators for measurements must cover a wide range of frequencies. This frequency range is the equivalent to the dynamic range of an amplifier. The frequency response of both types of oscillators is expressed as writing rate or slewing rate [recall (3-29)]. Data transmission oscillators are basically astable multivibrators. Typically, the center frequency is changed ± 7.5 or $\pm 15\%$ for telemetry and $\pm 40\%$ for magnetic-tape recorders.

Application 6-2   Sinusoidal Testing of High-Q Systems

A high-Q system requires a sinusoidal signal (either electrical or mechanical with low harmonic distortion for a test through a frequency range. Distortion becomes a problem when it is necessary to measure the response at an antiresonance or at a frequency wherein the system responds only slightly to the sinusoidal input. Suppose, for example, that the resonance was 60 dB

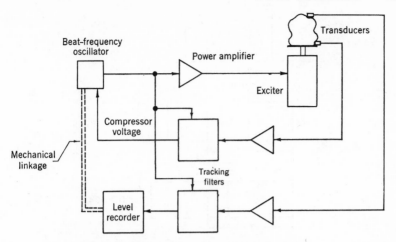

Figure 6-8   Sinusoidal testing of high-Q systems.

above the antiresonance and that the antiresonance occurred at one-third the frequency of the resonance. In this example an excitation system with 0.1 % of third-order harmonic distortion would excite the system at the anti-resonance and resonance equally. Thus it becomes necessary to filter the signal from the system. If testing is to be performed automatically as de-scribed in Application 6-1 a tracking filter as described in Section 2-4.4 is often used. See Figure 6-8.

6-3   Pulse Generators

It was shown in Section 2-2.1 that an impulse function can be used to illicit the equivalent information as that obtained from a harmonic sweep measurement using an oscillator. Because of the Fourier transform relation-ships between the impulse response of a system and its frequency and phase characteristics, overall system response can be determined by observing the system impulse pulse response on an oscilloscope.

The ideal pulse has an infinitely fast rise time as shown in Figure 6-9,

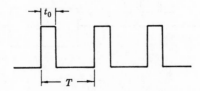

Figure 6-9   Ideal pulse train.

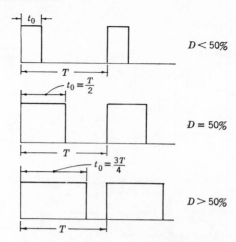

Figure 6-10   Pulse with various duty cycles.

where the pulse rate, $R = 1/T$ and the pulsewidth, t_0, are shown. When a pulse generator is used for testing a system, the pulsewidth must be short and the pulse rate must be long compared to the response time of the system. These requirements are necessary for the pulse to closely approximate an impulse function as discussed in Sections 1-4.2 and 1-4.3.

The duty cycle in percent, D, of a pulse train was defined in (4-16) as $D = (t_0/T) \times 100\%$. A pulse train with a duty cycle of less than 50%, 50%, and greater than 50% is shown in Figure 6-10.

Consider an astable multivibrator and bistable multivibrator as shown in Figure 6-11. Recall that an astable multivibrator may be synchronized from an external signal. The astable controls the period of the bistable. The period, or "on" time, of the bistable depends on external components which may be adjustable. Thus the period of the pulse is determined by adjusting internal components of the astable and the duration of the pulse is adjusted by changing external components of the bistable. Additionally, the astable

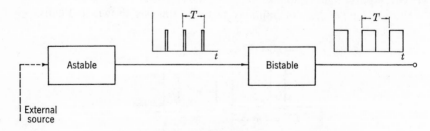

Figure 6-11   Functional diagram of a pulse generator.

or the bistable may be triggered by an external source such as an oscillator. In general, the frequency stability of a pulse generator is poor compared to an oscillator, which is clear if we consider the Q of both circuits. Thus if stable pulses are required, such as those used in timing, an external oscillator is used to trigger the pulse generator.

In order for a pulse generator to produce the ideal pulse shown in Figure 6-9 all frequencies from 0 to ∞ have to be passed through the pulse generator output circuitry. As was implied in Chapters 1 and 2 this is impossible. Thus the real pulse has a finite rise time and some fluctuations or overshoot due to the alteration of its high-frequency components. Also, if the duty cycle is long and the output of the generator is near maximum, the pulse may decay with time. This decay is called sag or droop. As can be seen from its rms value given in (4-27) and (4-28) a short pulse with a small duty cycle requires less power than a long pulse with a duty cycle approaching 50%.

Since the real pulse can only approach the ideal pulse the following definitions are commonly used for rise time, pulsewidth, and decay time. The rise time is defined as the time required for a pulse to rise from 10% of the final amplitude to 90% of the final amplitude (recall Section 2-2.5). The decay time is defined as the difference in time required for a pulse to decay from 90% of the final amplitude to 10% of the final amplitude. The pulsewidth is defined as the time required for the pulse to obtain 50% of the final amplitude during rise and decay. A typical real pulse is shown in Figure 6-12.

For pulse testing it is important that rise and decay times are significantly faster than the system or circuits to be tested. Any overshoot, ringing, or sag in the test pulse should be known, so as not to be confused with similar phenomena caused by the system under test. When a pulse is used for testing, the usual recorder is the oscilloscope. In general, the response of the circuit to both rise and decay are important. If the oscilloscope sweep initiates from the start of the pulse only the decay of the pulse may be seen. Furthermore, if there is a significant delay in the system the system response of the pulse may not be recorded. For these reasons most pulse generators have triggers for initiating the oscilloscope sweep. An adjustable delay network is also included in the pulse trigger to compensate for any delays in the system under test. This network delays the trigger pulse with respect to the test pulse. By adjusting the pulse-trigger delay, the pulse can be positioned at any convenient viewing position on the oscilloscope. (Recall Section 5-6.)

In order to demonstrate the physical behavior of circuits to a pulse input consider first a low-pass RLC circuit and then a high-pass RC circuit. The RLC circuit is shown in Figure 6-13. The pulse may be considered to be the sum of a step voltage of magnitude $+e_g$ at $t = 0^+$ and a step voltage of magnitude $-e_g$ at $t = t_p^+ (t_p > 0)$. Then if the response to a step e_g at $t = 0$ is $s_o(t)$, the output for $t > t_p$ is $s_o(t) - s_o(t - t_p)$, where it is seen that

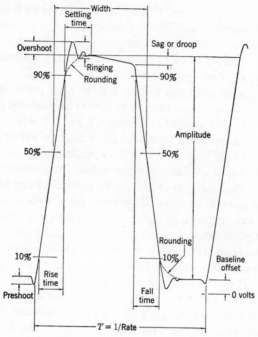

Figure 6-12   A real pulse and its primary characteristics.

t_p is the pulse duration. Thus only the unit step solution need be considered and these results are then suitably shifted in time to produce the response to a pulse.

It can be shown* that the response of the RLC to a step of $+e_g$ is the following.

$k = 1$:

$$\frac{s_o(t)}{e_g} = a\left[1 - \left(1 + \frac{t}{T}\right)e^{-t/T}\right]$$

(6-3a)

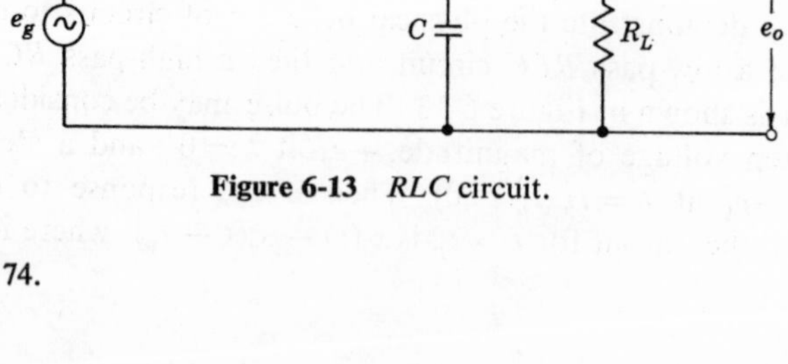

Figure 6-13 RLC circuit.

* See ref. 3, p. 74.

$k > 1$:

$$\frac{s_o(t)}{e_g} = a\left[1 - \frac{4k^2}{4k^2 - 1} e^{-t/2kT} + \frac{1}{4k^2 - 1} e^{-2kt/T}\right] \qquad (6\text{-}3b)$$

$k < 1$:

$$\frac{s_o(t)}{e_g} = a\left[1 - \left(\frac{k}{\sqrt{1 - k^2}} \sin \sqrt{1 - k^2}\, \frac{t}{T} + \cos \sqrt{1 - k^2}\, \frac{t}{T}\right) e^{-kt/T}\right]$$

$$(6\text{-}3c)$$

where $a = R_L/(R_g + R_L)$, $T = \sqrt{aLC}$, and $k = T(R_g/L + 1/R_LC)/2$. The case $k = 1$ is called the critically damped case, $k > 1$ the overdamped case, and $k < 1$ the underdamped case. The results given by (6-3) are shown in Figure 6-14. The transfer function for the system is given by

$$|H(\omega)|^2 = A_0^2[(\omega^2 T^2 - 1)^2 + (2k\omega T)^2]^{-1} \qquad (6\text{-}4)$$

where A_0 is a constant. [Recall (2-42).] Equation 6-4 is plotted in Figure 6-15.

It is seen that the departure from a vertical rise time is particularly noticeable if the pulse is especially lacking in high-frequency harmonics as in the overdamped ($k = 1$) case. The "corner" between the leading edge and the flat top of the pulse will be rounded for the same reason. If the pulse rises above the final value, as in the underdamped case, and then falls to the final value the output is said to ring. Overshoot indicates that some high-frequency harmonic components are present in excessive amounts as can be seen by examining Figure 6-15.

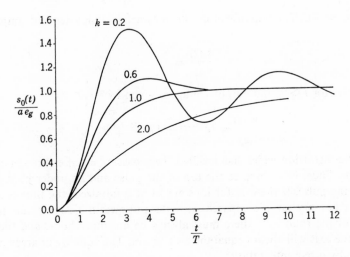

Figure 6-14 The rise-time response of the *RLC* circuit in Figure 6.13.

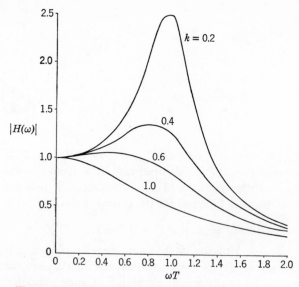

Figure 6-15 The transfer function of the *RLC* circuit in Figure 6-13 for various amounts of damping.

Now consider the response of a *RC* high-pass filter to a unit step. The transfer function was given by (2-51) as

$$|H(\omega)|^2 = \frac{\omega^2\tau_0^2}{1 + \omega^2\tau_0^2}$$

where $\tau_0 = RC$. The response of the *RC* network to a unit step with amplitude e_g is

$$\frac{s_o(t)}{e_g} = e^{-t/\tau_0}$$

Similarly, for $t > t_p$ the output voltage is given by

$$\frac{s_o(t)}{e_g} = (e^{-t_p/\tau_0} - 1)e^{-(t-t_p)/\tau_0}$$

Note the distortion which has resulted from passing a pulse through an *RC* network. There is a droop to the top of the pulse and an undershoot at the end of the pulse. If these distortions are to be minimized, the time constant τ_0 must be very large compared with the pulsewidth t_p. However, for all values of the ratio τ_0/t_p there must always be an undershoot, and the area below the axis will always equal the area above. The equality of areas can be verified by direct integration.

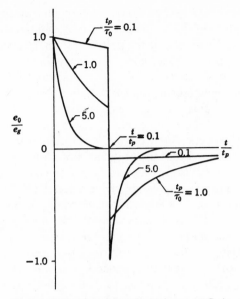

Figure 6-16 Response of a high-pass RC circuit to a pulse for several t_p/τ_0.

If the time constant is very large such that $\tau_0/t_p \gg 1$, there is only a slight droop to the output pulse and the undershoot is very small. If the time constant is very small $(\tau_0/t_p \ll 1)$, the output consists of a positive spike since this circuit acts as a differentiator (recall Section 2-3.2). We see that low-frequency distortion is indicated by a slope or general curvature of the top of the pulse, but it does not affect the rise time of the pulse. These various cases are shown in Figure 6-16.

Application 6-3   Measurement of Shock Wave Form and Spectra

Shock phenomena are encountered from explosions, impacts, earthquakes, supersonic motion, and other spasmodic releases of energy. The recording and measurement of shock wave forms impose stringent requirements upon the instrumentation employed. In general, a system to measure shock consists of a transducer, such as an accelerometer, microphone, or strain gauge, an amplifier, and a recorder. If the spectrum of the shock is to be found the recorder is usually a magnetic-tape recorder. The shock is recorded and then a tape loop of the shock is made and analyzed (see Section 1-4.5). If only the time record is necessary an oscilloscope is used. However, if an oscilloscope is used some provision must be made for triggering the oscilloscope before the shock occurs or the initial portion of the shock will not be recorded since it has been used to trigger the oscilloscope.

Referring to the Fourier spectrum of a rectangular pulse shown in Figure 1-7 it is seen that most of the energy of the pulse is contained in the frequency band from zero to $1/T$ where T is the pulse duration. In order to handle such a pulse the system must have an upper frequency limit equal to at least $1/T$. Not only the amplitude characteristics but also the phase characteristics of the measuring must be uniform (constant) up to at least $1/T$ in order to avoid distortion of the wave form. The phase characteristics of electronic amplifiers is usually reasonably uniform up to about one tenth of the cutoff frequency, so that an upper frequency limit of $10/T$ is adequate for most applications.

Recall from Section 6-3 that the peak value of a pulse decayed at the rate $1 - \exp(-t/\tau_0)$ where τ_0 is the time constant of the measurement system. In order that this quantity $1 - \exp(-t/\tau_0)$ be less than 0.05 (5% error), the term $\exp(-t/\tau_0)$ should be 0.95 or larger for the rectangular pulse. For this to be true, t/τ_0 should equal 0.05. In other words, the time constant, τ_0, of the measuring system should be 20 times the duration of the pulse.

From the foregoing discussion the following frequency response is required from a shock-measuring system for a general pulse wave form: low-frequency requirements; $\tau_0 \geqq 20T$, $f_L \leqq 0.008/T$. The high-frequency requirements are $f_h \geqq 10/T$. These limits are calculated for a rectangular wave form with a maximum of 5% low-frequency error and a tolerable rounding of corners due to high-frequency limitations. For practical wave forms a narrower frequency range may be acceptable. In particular, the low-frequency limit can be made considerably higher, if one is aware of the reason for the undershoot at the end of the pulse.

6-4 Random-Noise Generators

As a test signal, noise has an advantage over other test signals since it usually contains a broad band of frequencies. Evaluation of the output of a device under test gives not only frequency response, but also cross-modulation products. Another advantage of noise is that it offers random variations in amplitude. It can, therefore, be used to simulate natural occurring disturbances such as earth tremors, wave motion, vibration, and air turbulence.

Recall from Section 3-2 that a diode had white noise (shot noise) throughout a given frequency range. A typical noise-generating circuit uses two diodes as the noise source. The diodes are subjected to a dc current of such magnitude that the diodes are at the point on the noise characteristics which gives the highest noise voltage. The ac component of the noise voltage is fed to a summing amplifier and then to a buffer amplifier. Some noise generators also include a provision for switching a weighting network into the output of the generator. The weighting network slopes at -3 dB/octave (recall Application 2-5) and is useful in acoustic testing as discussed in Application 6-4.

The required bandwidth for uniform spectral density of the noise generator depends upon the use of the noise generator. For example, for acoustic testing in the audio frequency range the noise generator should have a uniform spectral density in the frequency range of 20 to 20 kHz. Additionally, it is important that random noise not be contaminated by hum related to the power-line frequency, by $1/f$ noise, which might arise in amplifiers, or by other disturbances. Periodic signals would correlate when the noise generator is used in a correlation analysis. Also, the presence of periodic signals or low-frequency semiconductor noise ($1/f$ noise) would alter the spectrum of the noise.

The amplitude probability density function from the random-noise generator should very closely follow the true Gaussian distribution out to $\pm 4\sigma$, where σ is the standard deviation. Such amplitudes must be supplied from the noise generator, without clipping, at the maximum rms voltage. For example if the rms value of the noise was 40 volts the maximum peak voltage required is 160 volts.

Application 6-4   Wide-Band Random-Vibration Testing

Wide-band random-vibration testing is used because the vibration producing mechanisms found in nature are more often of a random type than of a sinusoidal type and a random vibration test can be used to simulate the statistical character of certain vibration environments.

If an electrodynamic exciter, which is loaded with a test specimen, is connected to a random-noise generator with wide-band random noise of constant spectral density, the spectrum of the acceleration at the base of the exciter might look like the top curve shown in Figure 6-17. Further, the weight of the test specimen and the location of its resonant frequencies greatly influence the shape of this curve. Several methods have been developed to compensate electronically for these peaks and notches. One method consists of using a number of "peak-notch" equalizers, similar to the notch filter described in Section 2-4.6, in cascade. Each such equalizer would exhibit the inverse response of one resonance as measured on the vibration exciter base when the system is unequalized. By using as many equalizers as there are resonances it is possible to achieve a response similar to that given by the bottom curve shown in Figure 6-17. The curve shows the resulting vibration of the exciter base when both the frequency nonlinearities of the vibration exciter itself and those due to specimen resonances have been ideally compensated. However, the time spent adjusting for a particular test with this method is great.

A second equalization method is the so-called multiband equalization technique. This technique involves dividing the frequency range into a

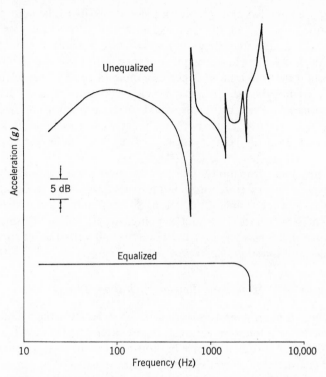

Figure 6-17   Example of ideal equalization of a wide-band random-vibration test.

number of discrete filter bands and adjusting the attenuation (or amplification) of each band individually. The adjustment can be performed manually or automatically by use of detectors and compressor amplifiers. The block diagram of an automatic multiband equalization system is shown in Figure 6-18. A disadvantage of multiband systems is that for a relatively accurate compensation of complex test specimens a large number of filters is required. However, one great adantage of using automatic equalization is the automatic correction for the effect of frequency response changes with change in test specimens.

Application 6-5   Measurement of Reverberation Time

One of the most important characteristics describing a room's acoustical quality is its reverberation time. When a sound source is suddenly switched on in a room, or the sound intensity of a source changes abruptly, there is a period during which the sounds from the previous instant still persist. This persistence is a function of the sound reflection (and absorption) properties

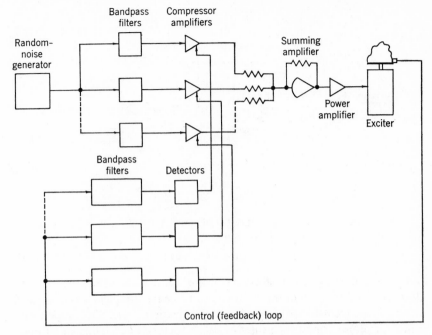

Figure 6-18 Automatic multiband random-vibration test system.

of the room. The reverberation time is defined as the period of time it takes from the moment a sound source placed in the room is switched off, until the sound level in the room has decreased to 0.001 of its original value. This represents a 60-dB decrease in the sound pressure level.

To measure the reverberation decay time, a system consisting of a sound source, a microphone with associated microphone amplifier, and a level recorder are used. Since it is important to determine the dependency of the reverberation time with frequency, either the sound-transmitting source or the sound-receiving system must be frequency selective. If the ambient noise level is high both must be frequency selective since the sound source must be at least 60 dB above the ambient noise level.

A system shown for the measurement of reverberation time in high ambient noise levels is shown in Figure 6-19. In this case the noise source is a random-noise generator which has been filtered through a set of $\frac{1}{3}$-octave filters. The filtering is used to obtain the maximum power from the power amplifier speaker combination in the frequency range of interest [recall (1-129a)]. A -3-dB/octave weighting is used with the random-noise generator to compensate for the 3-dB/octave increase in bandwidth of the octave filters (recall Application 2-5). The sound-receiving system consists of a microphone

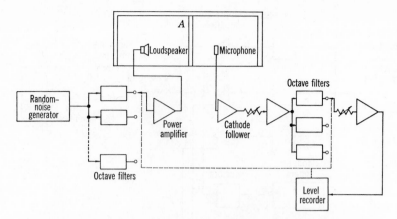

Figure 6-19 Measurement of reverberation time when partition *A* is removed and sound transmission loss when it is in place.

and microphone amplifier, a $\frac{1}{3}$-octave filter, and a level recorder. The two filter sets are both synchronized at the same center frequency and advanced by the level recorder. A second synchronization turns the random-noise generator off and on at the center of each $\frac{1}{3}$-octave filter. Thus an automatic reverbation decay time as a function of $\frac{1}{3}$-octave bands is obtained. In Figure 6-19 panel A is removed for this test.

Application 6-6 Measurement of Sound Insulation

The air-borne sound-transmission loss of a wall is defined as the ratio between the incident sound power (N_1) to the transmitted sound power (N_2). Expressed in decibels the transmission loss, TL, is TL $= 10 \log_{10} N_1/N_2$. In laboratory measurements the wall is normally built between two test rooms in such a way that the sound is transmitted through the wall test area (S) only. An expression for the transmission loss can thus be found from the sound pressure level in the transmitter and receiver rooms, the wall area, and the sound absorption of the receiver room. The sound power hitting the wall is, in this case, $N_1 = P_1^2 S/\rho c$ where $P_1^2/\rho c$ is the sound energy density in the transmitter room, P_1 is the rms sound pressure, ρ is the density of air, and c is the velocity of sound in air. Because the sound power transmitted into the receiver room per second must be equal to that absorbed by the surface of the room (absorption A), it can be found from the energy relation $N_2 = P_2^2 A/\rho c$. The expression normally used to determine the sound transmission loss of a wall, when measured in the laboratory is, therefore, TL $= L_1 - L_2 + 10 \log_{10} S/A$, where L_1 and L_2 are the sound pressure levels in the transmitter and receiver room, respectively, expressed in decibels.

The absorption A is found from the reverberation time of the receiving room (see Application 6–5) and is given by $A = 0.921\ Vd/c$ where V is the volume of the receiving room, c is the speed of sound, and d is the rate of decay of the sound in dB/sec.

The measurement system consists of a loud speaker, microphone, microphone amplifier, two octave filter sets, a level recorder, and a random-noise generator as shown in Figure 6-19. The center frequency of the octave filters are synchronized and controlled remotely from the level recorder. Wide-band random noise may, of course, also be used as a sound source, eliminating the use of one filter set. However, when the sound insulation in ordinary office buildings and dwellings is measured, the background noise level in the receiver room is often so high that a maximum excitation in the transmitter room is necessary.

Even though the use of noise bands and two or more loudspeaker units will sometimes produce a sound field which is fairly diffuse, it may be necessary to measure the sound pressure level both in the transmitter and in the receiver rooms at various points. The sound pressure level may be measured at each point one at a time or a switching system may be used to switch the microphones.

6-5 Summary

The transfer function of a system can be obtained using a sine wave, an impulse function, or white noise as an input. Simple oscillators are limited in continuous frequency range to approximately 1 decade. The beat-frequency oscillator does not have this restriction, and furthermore its output voltage may be controlled from an external voltage. When an oscillator is used to test a high-Q system the amount of harmonic distortion is an important specification because the harmonic components may excite other resonances in the system. If a precise frequency is required, a crystal oscillator is used.

Pulse generators are used for extremely rapid testing of systems. The entire system transient characteristics, amplitude as well as phase, may be displayed many times per second. It is important that the rise and decay times of the pulse train are significantly faster than the system or circuits to be tested. Additionally, any ringing or sag in the test pulse should be known, so as not to be confused with similar phenomena caused by the system under test.

Random noise evaluates the output of the device under a band of frequencies simultaneously and is often used to simulate actual environments. The amplitude probability density function from the random-noise generator should be Gaussian to $\pm 4\sigma$.

References

1. J. T. Broch, "The Application and Generation of Audio Frequency Random Noise," B & K Instruments, Cleveland, Ohio, Technical Review No. 2, 1961.
2. M. S. Ghausi, *Principles and Design of Linear Active Circuits*, McGraw-Hill Book Company, Inc., New York, 1965.
3. J. Millman and H. Taub, *Pulse, Digital and Switching Waveforms*, McGraw-Hill Book Company, Inc., New York, 1965.
4. J. D. Ryder, *Electronic Fundamentals and Applications*, Prentice-Hall, Inc., Englewood Cliffs, N.J., 1965.
5. *Typical Oscilloscope Circuitry*, Tektronix, Inc., Beaverton, Ore., 1966.

7

DIGITAL SYSTEMS—
COMPONENTS AND
TECHNIQUES

7-1 Introduction

Whenever scientific measurements are made, a digital number is the desired result. Most of the information in these measurements first appears in analog (continuous) form, and, therefore, must undergo a conversion or digitizing process. There are two main parts in the digitizing procedure. The first part is *sampling*, which is defining the points at which the data are observed. It seems reasonable to expect that there exists a minimum number of samples that are required to describe properly the significant information in the signal. The second part is the actual *quantization* of the signal. Quantizing of the signal is the approximation of the continuous signal by discrete levels (values). The quantizing procedure inherently introduces an error. However, both sampling and quantizing are governed by certain relations that permit one to minimize both error and redundancy of samples. These relations or theorems are discussed in detail in this chapter. In addition, the functions of digital logic elements are introduced, and their subsequent use in analog-to-digital converters, counters, shift registers, and digital delay lines are discussed. Furthermore, since digital information can be utilized easily by a digital computer, the organization and properties of small, general-purpose digital computers are introduced and discussed, with special regard to their functions as real-time devices.

7-2 Sampling Principles

7-2.1 Band-Limited Signals

The sampling principle specifies the least number of discrete values (samples) of an unknown function necessary for its complete and unambiguous definition. A restricted, but widely used, form of the sampling principle states:

233

If a signal that is a magnitude-time function is sampled instantaneously at regular intervals and at a rate slightly higher than twice the highest significant signal frequency, then the samples contain all of the information of the original signal.

This frequency is sometimes called the Nyquist frequency.

To prove the sampling theorem it is assumed that a signal $f(t)$, band-limited to B, has been sampled at uniform intervals of $T_0 = 1/2B$. It will first be shown that $f(t)$ may be reconstructed from these samples and that the actual reconstruction can be obtained by passing the sampled signal through a low-pass filter. The concept of band-limited function is given explicitly by the definition of the Fourier transform pair for a periodic signal. Recall (1-3a) which stated that

$$f(t) = \sum_{n=-\infty}^{\infty} C_n e^{jn\omega_0 t} \tag{7-1}$$

If the signal $f(t)$ is band-limited, (7-1) becomes

$$f(t) = \sum_{n=-N}^{N} C_n e^{jn\omega_0 t} \tag{7-2}$$

wherein N is a finite integer. In other words, a band-limited function contains no harmonics higher than the Nth.

Thus if $F(f)$ denotes the Fourier transform of $f(t)$

$$F(f) = \int_{-\infty}^{\infty} f(t)e^{-j2\pi ft} \, dt \tag{7-3}$$

then

$$f(t) = \int_{-B}^{B} F(f)e^{j2\pi ft} \, df \tag{7-4}$$

We now expand $F(f)$ in a Fourier series in the interval $-B < f < B$. Using (7-1) in the frequency domain yields

$$F(f) = \sum_{n=-\infty}^{\infty} C_n' e^{jn\pi f/B} \tag{7-5}$$

where the coefficients C_n' are given by [recall (1-4)]

$$C_n' = \frac{1}{2B} \int_{-B}^{B} F(f)e^{jn\pi f/B} \, df \tag{7-6}$$

Returning to (7-4) it is seen that the specific value of the band-limited function $f(t)$ at the regular spaced instants $t = n/2B$ is

$$f\left(\frac{n}{2B}\right) = \int_{-B}^{B} F(f)e^{jn\pi f/B} \, df \tag{7-7}$$

Comparing (7-6) and (7-7) it is evident that

$$f\left(\frac{n}{2B}\right) = 2BC'_{-n} \tag{7-8}$$

Thus the set of sample values of $f(t)$ given by $f(n/2B)$ (extending throughout the time domain) determines all the coefficients of the Fourier series expansion of the transform $F(f)$ in the interval $-B < f < B$ and hence completely specifies $F(f)$ itself within the interval of existence. Finally, from (7-4) the transform $F(f)$ uniquely determines the continuous time function $f(t)$.

The sampling theorem formalizes the fact that an ideally band-limited signal of bandwidth B cannot have independent values closer together, on the average, than $1/2B$. Thus sampling such a signal at a rate of at least $2B$ samples/sec effectively does preserve all the information contained in the original continuous wave form, and from the discrete samples it is possible to reconstruct its behavior between sample points.

The time function $f(t)$ is sampled periodically for a duration τ at intervals $1/2B$. If $\tau \ll 1/2B$, $f(t)$ may be assumed to be very nearly constant during the sampling time, then the Fourier transform of the individual sample $f(n/2B)$ is given by (7-3) as

$$F_n(f) = \int_{-\infty}^{\infty} f\left(\frac{n}{2B}\right) e^{-j2\pi ft}\, dt \approx \tau f\left(\frac{n}{2B}\right) e^{-jn\pi f/B}$$

From (1-61) it is seen that the Fourier transform of an impulse function displaced t_0 in time is given by

$$F(f) = \int_{-\infty}^{\infty} \delta(t - t_0) e^{-j2\pi ft}\, dt = e^{-j2\pi ft_0}$$

It follows therefore that $F_n(f)$ is just the Fourier transform of an impulse function of strength $\tau f(n/2B)$ located in time at $t = n/2B$. Thus by keeping $\tau \ll 1/2B$ a very small quantity the sample has effectively been approximated by an impulse whose magnitude is the same as the area $\tau f(n/2B)$.

To reconstruct the signal the following is noted. If (7-5) and (7-8) are substituted into (7-4) one obtains

$$f(t) = \sum_{n=-\infty}^{\infty} f\left(\frac{n}{2B}\right) \frac{\sin\left[2\pi B(t - n/2B)\right]}{2\pi B(t - n/2B)} \tag{7-9}$$

Equation 7-9 is plotted in Figure 7-1. This result shows that each sample $f(n/2B)$ is multiplied by the sampling function

$$\mathrm{Sa}(y) = \frac{\sin y}{y} \tag{7-10}$$

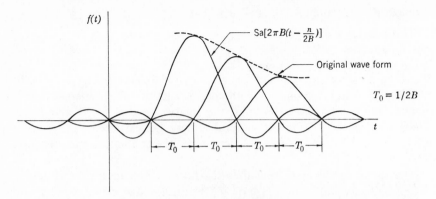

Figure 7-1   Reconstruction of original wave form from its sampled pulses.

Each of these sampling functions has its maximum at $t = n/2B$, the instant corresponding to the nth sample. Furthermore, at $t = k/2B$, $k \neq n$, the values of all the terms in the summation of (7-9) are zero; that is, at all the other sampling points $\text{Sa}(y) = 0$. In general, it is noted that at $t = n/2B \pm k/2B$, the term $f(n/2B)\,\text{Sa}(n/2B \pm k/2B)$ is virtually zero for $k > 2$. Thus each sampling function in the sum (7-9) materially influences the value of the sum only between $t = n/2B - 1/2B$ and $t = n/2B + 1/2B$ with the maximum occurring, as previously stated, at $n/2B$. Notice that (7-9) shows exactly how $f(t)$ should be reconstructed from the samples taken $1/2B$ apart.

Equation 2-14 gave the impulse response of an ideal low-pass filter as

$$K(\tau) = k' \frac{\sin\,[2\pi B(t - t_d)]}{2\pi B(t - t_d)}$$

where t_d is the delay of the filter. Thus from the preceding discussion, the fact that the sampled data are essentially impulse functions and considering (7-9), it is seen that the signal $f(t)$ can be obtained from the samples with an ideal low-pass filter followed by an amplifier of gain $1/k'$. Then the sum of impulses whose magnitudes are $f(n/2B)$ when passed through the filter would have an output of $f(t - t_d)$ wherein $f(t)$ is given by (7-9).

If this mathematical rigor is still not convincing the following heuristic remarks should help. Recall Section 2-2.5 which gave a relation stating that the rise time (or response time) of a low-pass filter is approximately equal to the reciprocal of twice its bandwidth, B. Thus if the pulses are any closer together in time, that is, have a shorter period, the filter will not be able to fully respond any faster than approximately $1/2B$. Hence if one were to sample as fast as the filter response time it should be sufficient to obtain all the information necessary since this is a minimum sampling rate with respect to all other signals whose frequencies are less than B.

7-2.2 Bandpass Signal

The sampling of a bandpass signal is now considered in order to establish the minimum sampling rate necessary for reconstructing the original signal from its samples. The sampling theorem of a bandpass function states:

When sampling a band of frequencies displaced from zero frequency, the minimum sampling rate f_s for a band of width B and highest inband frequency f_h is $2f_h/m$ where m is the largest integer not exceeding f_h/B. Thus

$$f_s = \frac{2f_n}{\text{INTEGER}[\leq f_n/B]} = \frac{1}{T_0}$$

where T_0 is the sampling interval. Obviously, not all higher rates are necessarily usable.

If we assume that $f_h = \alpha B$, $\alpha \geq 1$, then the sampling frequency can be written as

$$f_s = \frac{2\alpha B}{\text{INTEGER}[\leq \alpha]}$$

In this form it is easily seen that whenever α equals an integer, $f_s = 2B$ no matter how high the frequency range of the signal may be. When $1 < \alpha < 2$, f_s varies linearly between $2B$ and $4B$; when $2 < \alpha < 3$, f_s varies linearly between $2B$ and $3B$, and so on. This result is shown graphically in Figure 7-2.

The spectrum of the bandpass signal is assumed to occupy the frequency range $f_0 \leq |f| \leq f_0 + B = f_h$, wherein B is the bandwidth of the signal. Proceeding in the same manner as for the band-limited signal the relation analogous to (7-4) is

$$f(t) = \int_{-f_0-B}^{-f_0} F(f)e^{j2\pi ft}\, df + \int_{f_0}^{f_0+B} F(f)e^{j2\pi ft}\, df \qquad (7\text{-}11)$$

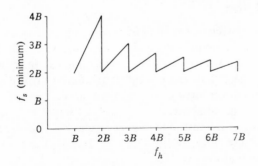

Figure 7-2 Minimum sampling frequency for a passband of width B.

If again $F(f)$ can be expressed in a Fourier series in the interval $f_0 < |f| < f_0 + B$ similar to that given by (7-5) and if we proceed in a manner analogous to that used to obtain (7-6) through (7-9) we find that

$$f(t) = \sum_{n=-\infty}^{\infty} f(nT_0) \frac{\sin [2\pi B(t - nT_0)/2]}{2\pi B(t - nT_0)/2}$$
$$\times \cos \left[2\pi \left(f_0 + \frac{B}{2} \right)(t - nT_0) \right] \qquad (7\text{-}12)$$

where T_0 is the sampling rate. Equation 7-12 shows exactly how $f(t)$ should be reconstructed from the samples taken T_0 apart.

Equation 2-15 gave the impulse response of an ideal bandpass filter as

$$K(\tau) = k_0'B \frac{\sin [2\pi B(\tau - t_d)/2]}{2\pi B(\tau - t_d)/2} \cos \left[2\pi \left(f_0 + \frac{B}{2} \right)(\tau - t_d) \right]$$

where t_d is again the delay of the filter. Thus comparing this result with (7-12) it is seen that the signal can be recovered by passing it through an ideal bandpass filter of bandwidth B.

The significance of these two sampling theorems is that they specify the least number of discrete values necessary to uniquely determine a continuous time function of limited bandwidth. As an example, consider a waveform that has been passed through a low-pass filter with a cutoff frequency of 5000 Hz. The wave should be sampled at a rate of 10,000 samples/sec to specify uniquely the band-limited wave without distortion. If another wave form has a frequency range from 5000 to 10,000 the minimum sampling rate needed is still 10,000 samples/sec. If, however, the frequency range would have been from 4999 to 9999, the minimum sampling rate now becomes 19,998 samples/sec. These two examples illustrate the limits of the sampling rate for the ideal case: $2B$ to $4B$. As a final remark it should be mentioned that the sampling need not be uniform, although this is usually the case.

7-3 Aliasing Errors

In the preceding section the minimum sampling rates are ascertained. If, however, a signal of frequencies higher than twice the sampling frequency is for some reason present in the signal to be sampled, an error is introduced in the sampled signal. This error is called an *aliasing* error and is one which does not occur in direct analog processing of signals.

Consider a signal composed of frequencies $0 < f < f_c$ where f_c is the highest frequency of interest. Any higher frequencies which are aliased with f are defined by $(2f_c \pm f)$, $(4f_c \pm f)$, $\ldots$, $(2nf_c \pm f)$ and so on. To prove this fact observe that for $t = 1/2f_c$

$$\cos 2\pi ft = \cos \left[\frac{2\pi(2nf_c \pm f)}{2f_c} \right] = \cos \left(\frac{\pi f}{f_c} \right)$$

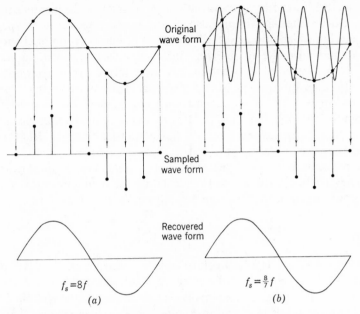

Figure 7-3 Aliasing.

Thus all signals with frequencies $2nf_c \pm f$ have the same cosine function as signals with frequencies f when they are sampled at points $1/2f_c$ apart. For example, if $f_c = 100$ Hz, then a signal at 170, 230, 370 Hz, and so on would be aliased with a signal of frequency 30 Hz. To illustrate the effect of aliasing consider Figure 7-3. In Figure 7-3a the sampling frequency is 8 times that of the sine wave. The set of sample values bears an obvious resemblance to the continuous wave form, and it is possible to reconstruct the original signal as accurately as desired. In Figure 7-3b, however, the sampling frequency is only $\frac{8}{7}$ that of the higher frequency sine wave. Now the set of sample values does not resemble the continuous wave form; in fact the higher frequency is deliberately chosen so that the samples appear to be derived from the lower frequency wave of Figure 7-3a, and the interpolation process will actually deliver this lower frequency signal output. Such downward spectral transposition of message power occurs whenever the minimum requirements of the sampling theorems are not satisfied. This phenomenon is, as previously mentioned, termed aliasing.

Aliasing is a potential danger in all sampled-data systems. It would be far preferable if the system simply failed to deliver any output in the continuous signal spectrum above the sampling frequency. Such, however, is not the case, and a false output component is developed which is indistinguishable from the real lower frequency variations, and which consequently cannot be eliminated by filtering without also distorting the true measured spectrum. The question which now arises is "what type of interpolation filter should be used to minimize the total error, including aliasing, when the sampled signal is reconstructed to a continuous wave form?" In other words what should the filter spectrum be when a given signal plus noise having a spectrum $S(f)$ is sampled and then reconstructed by passing the sampled pulses through the interpolation filter. In ref. 10 it has been shown that the optimum interpolation filter (see Section 2-8) has a transfer function $H(f)$ with an asymptotic rate of attenuation that is twice the frequency spectrum cutoff rate of the continuous signal being sampled and that at the Nyquist (sampling) frequency $H(f)$ is 6 dB down. In calculating the minimum error of such a filter it was assumed that the signal spectrum $S(f)$ is of the form

$$S(f) = \left[1 + \left(\frac{f}{f_0}\right)^{2k}\right]^{-1}$$

where the cutoff rate is $6k$ dB/octave and the spectrum bandwidth is defined by the -3-dB frequency of the filter, f_0. Notice that this is the spectrum of white noise through a maximally flat or Butterworth-type filter [given in (2-73)]. The result is shown in Figure 7-4. This figure shows that unless the rate of spectrum cutoff is quite high, sampling frequencies greatly in excess of twice the nominal (half power) bandwidth of the message spectrum are required to maintain the aliasing plus interpolation error within tolerable limits of, say, 1% rms. Conversely, if the spectra of the continuous signal are sharply restricted to a known cutoff frequency band, say a 60-dB/octave

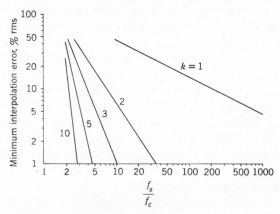

Figure 7-4 Minimum interpolation error versus normalized sampling frequency.

($k = 10$) cutoff filter, then the sampling frequency of about 3 times the nominal signal bandwidth suffices to ensure 1 % rms aliasing plus interpolation error.

7-4 Binary Codes

7-4.1 Straight Binary

The most common form used to represent digital numbers electronically is the binary system which involves only two digits, ordinarily designated as 0 and 1. The reason for this nearly universal preference for binary arithmetic in digital coding is that it requires only a simple yes-no decision on the part of the electronic circuitry concerned. It also happens that the binary system offers the greatest advantages in reducing the effect of noise.

Binary arithmetic uses the symbols 0 and 1 to designate whether or not a power of 2, corresponding to a digit position in a binary word, should be included in a sum which yields the equivalent decimal number. Any decimal number is expressible in binary terms by simply using a word long enough; that is, 2^n. For example, 11 binary digits can be used to express any decimal number up to $2^{11} - 1 = 2047$. (There are, however, 2048 decimal numbers if we include zero.) Thus if a_k stands for either 0 or 1 the binary notation with n digits, $a_1, a_2, \ldots, a_n$, represents the number

$$a_1 2^0 + a_2 2^1 + a_3 2^2 + \cdots + a_n 2^{n-1} = \sum_{i=1}^{n} a_i 2^{i-1} \qquad (7\text{-}13)$$

Thus the binary number 011 represents the number $1 + 2 + 0$. Notice that binary numbers are written backwards; that is, $a_n a_{n-1} \cdots a_1$. Table 7-1 lists the binary and the decimal equivalents for a six-digit ($n = 6$) system. The highest number in this system is $2^6 - 1 = 63$.

Table 7-1 Straight Binary Coding

Decimal Number	Binary Number
0	000000
1	000001
2	000010
.	.
.	.
.	.
10	001010
.	.
.	.
.	.
62	111110
63	111111

Table 7-2

(a)		(b)	
Decimal Values	Straight Binary	Decimal Values	Offset Binary
0 to Δ	000	-4Δ to -3Δ	000
Δ to 2Δ	001	-3Δ to -2Δ	001
2Δ to 3Δ	010	-2Δ to $-\Delta$	010
3Δ to 4Δ	011	$-\Delta$ to 0	011
4Δ to 5Δ	100	0 to Δ	100
5Δ to 6Δ	101	Δ to 2Δ	101
6Δ to 7Δ	110	2Δ to 3Δ	110
7Δ to 8Δ	111	3Δ to 4Δ	111

If the numbers being considered are both positive and negative the above results cannot be used without upsetting the natural order of the binary numbers. To overcome this difficulty an *offset* binary code is used. In the offset binary code it is common for the quantity of binary numbers assigned to negative values to be equal to the quantity of binary numbers assigned positive values. The result of such an assignment is that the first bit in the binary numbers is a 0 for negative values and a 1 for positive values. This bit, therefore, indicates the polarity of the value and in such cases is referred to as the "polarity" or "sign" bit.

To illustrate the offset binary code consider a three-bit code $(n = 3)$ which can record the numbers 0 to A. In straight binary the values 0, Δ, $2\Delta, \ldots, A$, where $\Delta = A/2^3$, would be assigned the binary values as shown in Table 7-2a. With an offset binary code, the assignment would be that shown in Table 7-2b.

7-4.2 Binary-Coded Decimal (BCD)

From the preceding section we see that the straight binary system requires many more digits to express very large decimal numbers than are in the original decimal number. The binary-coded decimal (BCD) system can be employed to overcome this difficulty. The system uses a binary-coded number to express each decimal digit in some form of binary notation. For example, the number 1024 can be written in BCD notation as

$$\begin{matrix} 0001 & 0000 & 0010 & 0100 \\ 1 & 0 & 2 & 4 \end{matrix}$$

Although this BCD notation requires 16 binary digits to represent a four-digit decimal number, it is simple to code, easy to read, and the 16 binary digits accommodate decimal numbers up to 9999.

We see that a minimum of four binary bits is required to represent a decimal digit or decade. The most common and straightforward method of

BCD notation is the 8, 4, 2, 1 code, which comes directly from (7-13) for $n = 4$, and is called the "standard" or "natural" BCD code. Another BCD code frequently used is the 2, 4, 2, 1 code. Notice that in this code the decimal digits 2 through 7 are not necessarily unique. For example, either 1000 or 0010 may be used to indicate the decimal value 2. The advantages and disadvantages of these codes are most often important only if subsequent arithmetic operations are to be performed with the particular code. More frequently the BCD output goes to a digit computer and hence is converted to a binary format acceptable to that particular computer.

7-4.3 Gray (Cyclic) Code

The Gray code is sometimes called a reflected code; that is, one in which only one bit position changes every time the number is increased one unit. Since the Gray code does not lend itself to computation, a conversion to a straight binary code is frequently employed. The binary code and its equivalent Gray code is given in Table 7-3 for the decimal numbers 0 through 15.

This table is obtained from the relation

$$B_n = R_n$$
$$B_j = B_{j+1}\bar{R}_j + \bar{B}_{j+1}R_j \qquad n \neq j$$

where R_j is the jth bit in the reflected code for a given number, B_j the corresponding bit in the binary code for the same number, and the nth digit the

Table 7-3 Gray Code

Decimal	Binary				Gray			
0	0	0	0	0	0	0	0	0
1	0	0	0	1	0	0	0	1
2	0	0	1	0	0	0	1	1
3	0	0	1	1	0	0	1	0
4	0	1	0	0	0	1	1	0
5	0	1	0	1	0	1	1	1
6	0	1	1	0	0	1	0	1
7	0	1	1	1	0	1	0	0
8	1	0	0	0	1	1	0	0
9	1	0	0	1	1	1	0	1
10	1	0	1	0	1	1	1	1
11	1	0	1	1	1	1	1	0
12	1	1	0	0	1	0	1	0
13	1	1	0	1	1	0	1	1
14	1	1	1	0	1	0	0	1
15	1	1	1	1	1	0	0	0

Table 7-4

Decimal Values	1's Complement
-4Δ to -3Δ	100
-3Δ to -2Δ	101
-2Δ to $-\Delta$	110
$-\Delta$ to 0Δ	111
0 to Δ	000
Δ to 2Δ	001
2Δ to 3Δ	010
3Δ to 4Δ	011

1's complement

most significant one in both codes (i.e., the first 1 reading from left to right). The bar above R_j and B_j denotes the opposite state of the R_j and B_j without the bar. Thus if $R_j = 1$, then $\bar{R}_j = 0$.

7-4.4 Complement Codes

In some applications, the binary number zero must be assigned to a quantity whose magnitude is zero, while still maintaining the remaining binary numbers in a workable order. The straight binary code fails to do this when both positive and negative values are to be considered. For these cases the offset binary can be used or the 1's complement code. The 1's complement code is created by taking the straight binary numbers in their natural, numerical order and inverting each bit (1 becomes 0, 0 becomes 1). Using the three-bit system given in Table 7-2b, Table 7-4 gives an illustration of the 1's complement. Notice that this table is readily obtainable from Table 7-2b by simply inverting the first bit of each offset binary number.

In the 1's complement code 000 is not at the center of the -4Δ to $+4\Delta$ range of A. To position the binary code such that 000 is at the center another type of complement code, called the 2's complement code, is used. The 2's complement of a binary number is obtained by adding one to the 1's

Table 7-5

Decimal Value	2's Complement
$-5\Delta'$ to $-4\Delta'$	100
$-4\Delta'$ to $-3\Delta'$	101
$-3\Delta'$ to $-2\Delta'$	110
$-2\Delta'$ to $-\Delta'$	111
$-\Delta'$ to $+\Delta'$	000
Δ' to $2\Delta'$	001
$2\Delta'$ to $3\Delta'$	010
$3\Delta'$ to $4\Delta'$	011

2's complement

complement. For example, the 1's complement of 010 is 101 and the 2's complement of 010 is 101 plus 001 or 110. The 2's complement assignment results in an unsymmetrical total amplitude range. The magnitude of this asymmetry decreases for codes wherein n is large and is not considered a serious disadvantage. The 2's complement is shown in Table 7-5.

7-4.5 Series/Parallel Binary Information

Most binary logic systems operate as synchronous systems wherein all operations are performed during a definite constant interval of time. To achieve this synchronism, a continuous sequence of pulses of good wave shape whose frequency is usually established by a crystal oscillator is used. This stable oscillator determines the basic rate at which the system operates. In addition, these pulses are distributed to all parts of the system where they are used to maintain the timing of the system.

In a synchronous system a number is represented by a time sequence of pulses. The pulses (or absence of pulses) occur serially, one after another, and the information (number) conveyed by this pulse sequence may be transmitted over a single communication link (e.g., a pair of wires). This

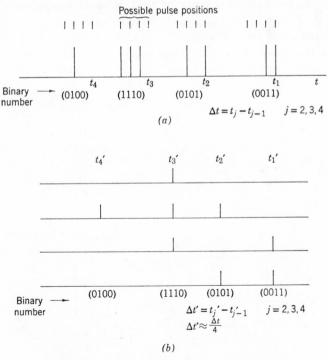

Figure 7-5 (*a*) Serial output and (*b*) parallel output.

mode of representing information is described by the word *serial*. In the serial mode the time required to transmit the information is the duration T of a pulse interval multiplied by the number of bits in the information.

When these pulses are transmitted over n wires, each wire carrying a 0 or 1 corresponding to a specific a_j of (7-13), the mode of transmission is called *parallel*. In the parallel mode the time required to transmit the information is just the duration of the pulse. However, the wiring of this type of a system becomes more complex. These two systems are illustrated in Figure 7-5 for a four-bit system ($n = 4$).

7-5 Quantization

Quantization is the approximate representation of a continuously varying wave form by a set of discrete values, each value being assigned a discrete number. It inherently introduces an initial error in the amplitude of the samples sometimes called quantization noise. But once the signal is in a quantized state, it can be relayed many times without further loss in quality, provided that only the added noise in the signal is not too great to prevent recognition of the particular level that each given signal is intended to represent. The quantizing noise can be reduced to an acceptable minimum by choosing the step or level separation fine enough.

In anticipation of the discussion to follow in Section 7-7 on analog-to-digital converters, we consider a voltage range of 0 to $+A$. Then as shown in Figure 7-6 the entire range is divided into $2^n - 1$ equal parts, where n is

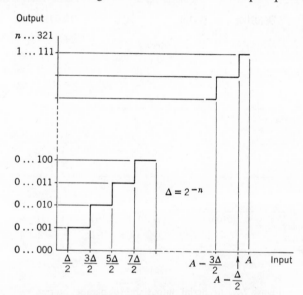

Figure 7-6 Ideally quantized interval 0 to A.

the number of bits, and each equal part or quantization level is 2^{-n} in magnitude. The ordinate would then correspond to the n-bit binary number. We see that for a 10-bit binary system ($n = 10$) and an amplitude of 10 volts ($A = 10$) there are 1023 equal levels, each of 9.765 mV. Thus this particular quantization is able to predict the continuous amplitude to one part in 2046 or approximately 0.05%. The reason that the accuracy is one part in 2046 rather than one part in 1023 ($2^{10} - 1$) is that the signal will incur its maximum error when it is midway between levels; any other position will have less error.

From the preceding discussion we see that there are two quantities that are useful in describing the resolving power of an analog-to-digital converter: least significant bit (LSB) and most significant bit (MSB). The most significant bit is the first bit (reading from left to right) of a binary number. Thus the MSB of the binary number 10001 is the first 1 which corresponds to weight 1 multiplying 2^4 [recall (7-13)]. The LSB is defined as the ratio of the total input amplitude range, A, divided by 2^n; thus

$$LSB = \frac{A}{2^n} \tag{7-14}$$

where n is the number of bits contained in the binary numbers that are assigned to the input signal. It should be pointed out that (7-14) is valid for straight binary codes but for offset binary, 2's complement and sign plus magnitude codes (7-14) must be multiplied by 2. Thus LSB $= A \times 2^{1-n}$. The

Table 7-6

Number of Bits (n)	MSB 2^n	LSB 2^{-n}	% Max Error ($\frac{1}{2}$ LSB $\times$ 100%)
0	1	1.0	50
1	2	0.5	25
2	4	0.25	12.5
3	8	0.125	6.25
4	16	0.0625	3.125
5	32	0.03125	1.563
6	64	0.015625	0.781
7	128	0.0078125	0.496
8	256	0.0039606	0.195
9	512	0.0019531	0.0977
10	1024	0.00097656	0.0488
11	2048	0.00048828	0.0244
12	4096	0.00024414	0.0122
13	8192	0.00012207	0.0061
14	16384	0.00006104	0.0030
15	32768	0.00003052	0.0015
16	65536	0.00001525	0.0007

reason for the factor of 2 is that the MSB gives only sign information and not magnitude information.

We note again that the quantizing (or digital) error has a *maximum* value equal to $\pm\frac{1}{2}$ LSB, which in the previous example was $\frac{1}{2046} \approx 0.05\%$. All other magnitudes between 0 and $\pm\frac{1}{2}$ LSB are equally likely. It can be shown that the *average* quantizing error is $\frac{1}{4}$ LSB and the rms value is $1/\sqrt{12}$ LSB $= 0.2886$ LSB. A table of values of the MSB and LSB is given in Table 7-6 as a function of n, the number of bits.

7-6   Logic Circuits

7-6.1   Introduction

The apparently complex logical operations performed by even the most sophisticated digital instruments are based on a relatively simple system of symbolic switching logic and implemented by repeated use of a few basic logic circuits. These basic logic functions and several of their representative applications are discussed in this section.

7-6.2   Logic Gates

OR Gate.   An OR gate has two or more inputs and a single output and operates according to the following definition. *The output of the OR assumes the 1 state if one or more inputs assume the 1 state.* The n inputs to a logic circuit will be designated by A, B, . . . , N and the output by Y. It is to be understood that each of these symbols may assume one of two possible values, either 0 or 1. The symbol for the OR circuit is given in Figure 7-7a. The *truth table* is shown in Figure 7-7b, which contains a tabulation of all possible input values and their corresponding outputs for the OR gate. It should be clear that the two-input truth table of Figure 7-7b is equivalent to the above definition of the OR operation.

AND Gate.   An AND gate has two or more inputs and a single output, and it operates in accordance with the following definition. *The output of an*

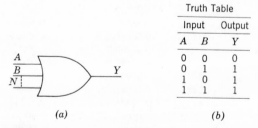

	Truth Table	
Input		Output
A	B	Y
0	0	0
0	1	1
1	0	1
1	1	1

(a) (b)

Figure 7-7   Logic symbol (a) and truth table (b) for an OR gate.

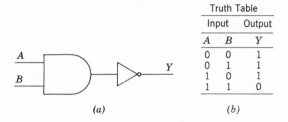

Truth Table

Input		Output
A	*B*	*Y*
0	0	1
0	1	1
1	0	1
1	1	0

(a)   (b)

Figure 7-8 Logic symbol (a) and truth table (b) for an AND gate.

AND assumes the 1 *state if and only if all the inputs assume the* 1 *state.* The AND circuit and its truth table is given in Figure 7-8.

NOT Circuit. The NOT circuit has a single input and a single output and performs the operation of *logic negation* in accordance with the following definition. *The output of a NOT circuit takes on the* 1 *state if and only if the input does not take on the* 1 *state.* The symbol to indicate logic negation is shown in Figure 7-9. A NOT circuit is sometimes called an *inverter*.

INHIBIT Operation. A NOT circuit preceding one terminal (*N*) of an AND gate acts as an inhibitor. This leads to the logical statement; if $A = 1$, $B = 1, \ldots, M = 1$, then $Y = 1$ provided that $N = 0$. However, if $N = 1$, then the coincidence of $A, B, \ldots, M$ is inhibited, and $Y = 0$. The logical symbol and the truth table are shown in Figure 7-10.

NAND and NOR Gates. The preceding basic circuits can be combined to produce other circuits. For example a NOT circuit when combined with an AND gate produces a NOT-AND or NAND gate. When a NOT circuit is combined with an OR gate it produces a NOT-OR or NOR gate. The logic symbols and their respective truth tables are given in Figures 7-11 and 7-12, respectively.

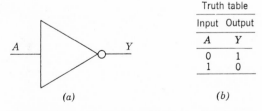

Truth table

Input	Output
A	*Y*
0	1
1	0

(a)   (b)

Figure 7-9 Logic negation (NOT) symbol (a) and its truth table (b).

Truth table			
Input			Output
A	B	N	Y
0	0	0	0
0	1	0	0
1	0	0	0
1	1	0	1
0	0	1	0
0	1	1	0
1	0	1	0
1	1	1	0

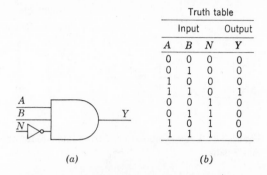

(a)   (b)

Figure 7-10   Logic symbol (a) and truth table (b) for an AND gate with an inhibitor at terminal N.

FLIP-FLOP. In addition to the basic circuits just described another important circuit is the flip-flop or binary circuit. A flip-flop consists of two NOT circuits interconnected in the manner shown in Figure 7-13. The most important property of the flip-flop is that, because of the interconnection, the circuit may persist indefinitely in a state in which one device (say $Q1$) is ON while the other ($Q2$) is OFF. A second stable state of the FLIP-FLOP is one in which the rules of the two devices are interchanged so that $Q1$ is OFF and $Q2$ is ON. Since a flip-flop has two stable states it may be used to *store* one bit of information.

From Figure 7-13a it is seen that the two stable states are either $Y = 1$ and $\bar{Y} = 0$ or $Y = 0$ and $\bar{Y} = 1$. To show this let the output Y of one NOT circuit be 1 so that the input A_2 to the second NOT circuit is also 1. The second NOT circuit then has the state 0 at its output Y, and thus at the input A_1 to the first gate. This result is consistent with the original assumption that the first NOT gate had a 1 as its output.

A flip-flop is represented in block form in Figure 7-13b where the three input terminals are indicated by $S =$ set, $R =$ reset, and $T =$ trigger. An

Truth table		
Input		Output
A	B	Y
0	0	0
0	1	0
1	0	0
1	1	1

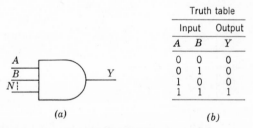

(a)   (b)

Figure 7-11   Logic symbol for a NAND gate (a) and its truth table (b).

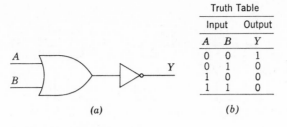

Truth Table

Input		Output
A	B	Y
0	0	1
0	1	0
1	0	0
1	1	0

(a)   (b)

Figure 7-12   Logic symbol for a NOR gate (a) and its truth table (b).

excitation (state 1) of the *set* input causes the flip-flop to establish itself in the state $Y = 1$. If the binary is already in that state the excitation has no effect. A signal at the *reset* input causes the flip-flop to establish itself in the state $Y = 0$. If the binary is already in that state the excitation has no effect. A triggering signal applied to the T input causes the flip-flop to change its state regardless of the existing state of the binary. Thus each successive excitation applied to T causes a transfer, and T is referred to as the *toggle* input. This type of excitation is called symmetrical triggering and is used in binary counters and other applications. Unsymmetrical triggering through the S or R input is also useful in logic applications.

7-6.3   Shift Register

Consider first an application of unsymmetrical triggering as shown in Figure 7-14a. This is the block diagram of a shift register. The input, shown in Figure 7-14b, consists of a train of pulses which is to be stored in the register. The reset or shift-pulse line is excited by a continuous train of pulses

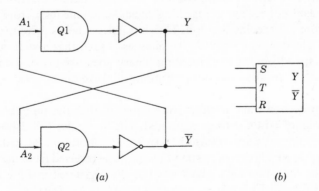

(a)   (b)

Figure 7-13   A flip-flop assembled from two NOT circuits (a) and its logic symbol (b).

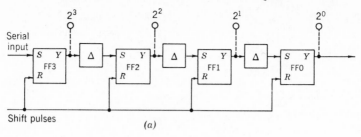

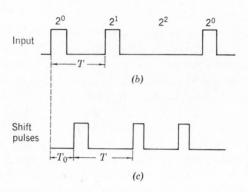

Figure 7-14 (*a*) Shift register, (*b*) typical input pulse train, and (*c*) the shift pulses ($\Delta \ll T$).

which are timed to occur nominally midway between the pulses of the input number. The delay sections have a delay Δ much smaller than the time interval between pulses and are required to ensure that an individual binary (flip-flop) shall not receive a triggering signal simultaneously from the shift line and from a preceding flip-flop. The shift pulses always drive the binaries to state 0. The coupling between the binaries is such that a succeeding flip-flop will respond only if the preceding binary goes from state 1 to state 0. The pulse which results from this transition will drive the succeeding flip-flop to state 1.

To illustrate how this network works assume that the number 1011, as shown in Figure 7-14*b*, is to be registered. The first pulse (2^0) drives FF3 to state 1. The shift pulse now returns FF3 to 0, and a short time later (depending on the delay Δ) FF2 is driven to state 1 by the pulse received from the previous binary. The first digit (2^0), which was initially registered in FF3, has been shifted to FF2 and FF3 has been cleared (returned to 0) so that it may now register the next pulse (2^1). It is seen that in this process of registering and shifting to make room for the next pulse, the input number will eventually become installed in the register. Of course, the shift pulses must cease at the moment the number has been registered.

To read the register serially at the output Y of FF0 it is necessary to apply four shift pulses. In response to these shift pulses the original number will appear at the output Y of FF0. Note that the shift-out rate may be greater or smaller than the original pulse repetition rate. Hence here is a method for effectively changing the spacing of a pulse sequence, a process referred to as *buffering*. If the output of the shift register is to be taken in a parallel fashion there would be, in this example, four additional wires added at the output Y of each flip-flop. These are shown by the dotted lines in Figure 7-14*a*.

A shift register may also function as a *digital delay line*. Thus the input pulse train appears at the output of an n-stage register delayed by a time equal to $T_0 + (n - 1)T$, where T is the interval between shift pulses and T_0 is the time between the first bit and the shift pulse (see Figure 7-14*c*).

7-6.4 Counters

Counting is achieved by a circuit that takes note of the number of input pulses and stores the result. If a counter adds each arriving pulse, it is known as a forward or up counter. If a counter can be set initially to a certain number and each input pulse subtracts from that number, it is known as a reverse or down counter. If a counter is provided with two input terminals so that it adds for pulses arriving at one input and subtracts for pulses arriving at the other input, it is known as a reversible or up-down counter.

Counters may also be classified according to the manner in which they store the number of counts. If the stored number is a binary number, the counter is a binary counter; if the stored number is a decimal number, the counter is a decimal counter. A decimal counter usually consists of a number of decimal-digit counters. If the decimal digit is stored in a binary code, the counter is a binary-coded-decimal (BCD) digit counter. Counters can also be designed to express the count in a special sequence of binary numbers. Such a counter may appear in a digital computer as a program counter.

An illustration of a simple counter is shown in the binary counter in Figure 7-15 (which is also an example of symmetrical triggering). The output Y of one flip-flop in the chain is coupled to the trigger T of the next flip-flop. Recall that the trigger input terminal is a point at which each successive signal will reverse the state of the flip-flop (or binary). Consider the case when 16 successive triggering pulses are applied to T of FF0. The first external pulse applied to FF0 causes the binary to make a transition from state 0 to state 1. Each of the other flip-flops makes a transition *only* when the preceding flip-flops remain at state 0. The second external pulse causes FF0 to return from state 1 to state 0. Hence FF1 now responds and goes from state 0 to state 1; the other two still remain at state 0. The process is easily continued.

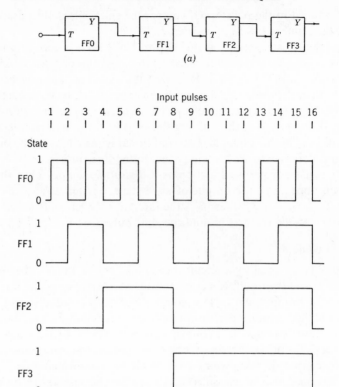

Figure 7-15 (*a*) A chain of flip-flop circuits used for counting and (*b*) the associated wave-form chart.

The results of all 16 pulses on the states of the four flip-flops are given in Figure 7-15*b*. It is seen that the ordered array of states 0 and 1 in any row in Figure 7-13 is precisely the binary representation of the number of input pulses given by the decimal numbers 0 to 15. Thus the chain of flip-flops counts in the binary system. A chain of n flip-flops or binaries will count up to the number $2^n - 1$ before resetting itself into the original state.

To read the count of a binary chain the state of each individual binary in the chain must be determined. A very rudimentary indicator may be used for this purpose since only the circuit that has the voltage need be recorded. Usually a small neon bulb is connected between the output Y and the ground. When the Y voltage is low, there is not enough voltage to make the tube glow, but it will glow when the voltage is high (state 1). The neon bulbs connected across the flip-flops FF0, FF1, FF2, and FF3 could be assigned the values 1, 1, 1, and 1 or 1, 2, 4, and 8, respectively. If the latter are used, it is only necessary to add the numbers assigned to those neon bulbs that are lit to determine the count.

7-7   Analog-to-Digital Conversion (Digitizing, Encoding)

7-7.1   Ramp-Type Converter

One method of converting an analog voltage to digital form is to compare it to a linearly varying sawtooth wave known as a ramp. As shown in Figure 7-16, the comparison requires the use of two voltage comparators. A comparator generates a 0 or 1 state depending on whether the analog input signal is less than or greater than the reference signal, respectively. One comparator produces an output at the instant the ramp voltage passes through zero or ground potential. This output opens the gate circuit and permits pulses at a crystal-controlled frequency to feed a binary counter. The other comparator produces an output at the instant the ramp voltage becomes equal to the analog voltage to be digitized. Since the ramp voltage builds up linearly, the time required for the ramp to reach the value of the analog voltage is proportional to the magnitude of the analog voltage. The binary number stored in the counter after the gate closes is, therefore, a digital representation of the value of the analog input. A readout pulse transfers the binary number

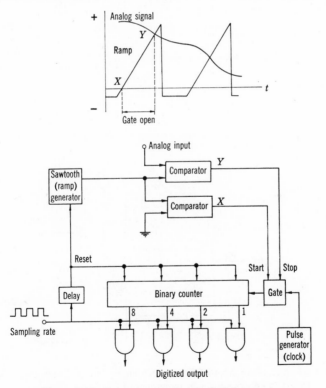

Figure 7-16   Ramp-type analog-to-digital converter.

out of the counter, which is reset. The same reset pulse also triggers the sawtooth generator to initiate another ramp.

A major advantage of this conversion method is simplicity and economy. However, the method has not gained a reputation for suitability to high-accuracy conversion and the attainable sampling rate is low in comparison with the other conversion methods to follow. To determine the maximum conversion C, we use the following relation: $C = f/2^n$, where f is the clock frequency and n is the number of bits. Thus for a 10-MHz ($f = 10^7$) clock frequency and 0.1% accuracy ($n = 10$; $2^{-10} \simeq 0.001$) the maximum conversion rate is 10,000 conversions/sec. This technique, however, is widely used in digital voltmeter applications (see Section 7-11.3) where the time required for conversion would be no problem. With this type of conversion technique the resulting digital number represents the analog signal at

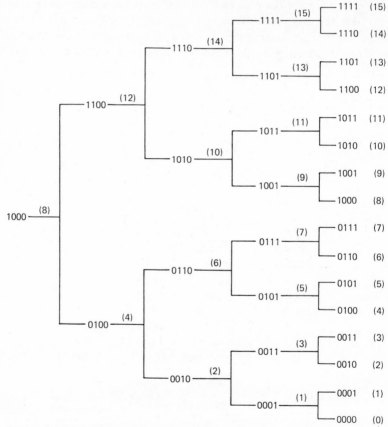

Figure 7-17 Successive approximation technique—binary fractional scale for $n = 4$. (Decimal numbers appear in parentheses.)

the time the conversion process is completed, which, in general, cannot be predicted in advance and may prove a disadvantage in some applications.

7-7.2 Successive-Approximation-Type Converter

Another method which is used to convert analog signals into digital words is the successive-approximation technique. This technique involves a feedback and comparison method to generate a digital output word equivalent to the amplitude of the analog input voltage. In the first approximation, a feedback voltage is generated which is equal to half the full input range of the converter. The feedback voltage is compared with the analog input voltage. If the feedback voltage is greater, a signal is generated by the comparator which removes the half-range feedback. In the second approximation a new feedback voltage which is equal to $\frac{1}{4}$ the full input range of the converter is superimposed on the result of the first approximation. If the half-range feedback voltage is less than the analog input voltage, the combined signal at the second approximation produces a new feedback voltage which is $\frac{3}{4}$ the full input range of the converter. Each approximation cycle superimposes a binary fractional-scale trial voltage on the result of the previous comparison. The result of each comparison is actually the decision to keep or delete each bit in a binary-coded representation of the full input range. This successive feedback-and-comparison technique is continued until the least significant bit of the digital output word is generated. An illustration of this binary fractional scale is shown in Figure 7-17 for a four-bit word and for the general case in Figure 7-6.

7-7.3 Cyclic Converter

A third type of analog-to-digital converter is the cyclic converter. A typical converter stage of a cyclic converter is shown in Figure 7-18. The input to the resistive adder network is either the analog input or the "residue" from the previous state. The resistive adder network is used to shift the input voltage level so that one-half the maximum voltage level is 0 volts to the decision network. The output of the adder network is then passed through an amplifier having a gain of -2. The output of the amplifier goes

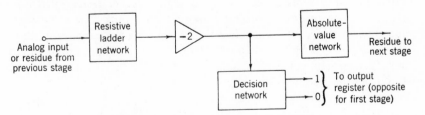

Figure 7-18 Typical state of a cyclic converter.

to the decision network which senses the polarity of the signal at that point and makes the output 1 if positive and 0 if negative. The first stages, however, operates inversely from the remaining stages; that is, a positive level will produce a 0 and vice versa. Also following the amplifier is an absolute-value network which takes the absolute value of the voltage of the "residue" and applies it to the next converter stage.

To illustrate the operation of the converter stage assume an input voltage of V_i volts. Let the maximum voltage of the system be V_m volts. Then the output of the resistive adder network and amplifier is $V_0 = -2(V_i - V_m/2)$. Depending on whether V_0 is positive or negative the decision network will give either a 1 or 0. The input to the next stage will be $|V_0|$ and the process is repeated.

The word generated by the cyclic converter is in Gray code rather than in conventional binary code. It is recalled that the characteristic feature of the Gray code is that only one bit changes at a time as the input signal varies and, therefore, the data word can be read at any time with only one bit of uncertainty. Since the results of the decision at each step in the conversion sequence are independent of the results of the preceding steps, if the analog input varies more rapidly than the converter can track, the converter will lag, but the data word will not be in error. Even though the output may not represent the analog voltage at the exact instant of sampling, it will correspond to the level which was occupied just prior to the sampling. It is this feature that eliminates the need for a sample-and-hold system for most applications.

7-7.4 Practical Considerations and Limitations of Analog-to-Digital Converters

Perhaps the single most important property of an A/D converter is how fast it can compare each reference voltage total to the voltage that one wishes to measure. In order to accomplish this comparison, the signal has to be switched in, compared (digitized), and settled out during the interval known as the bit-rate period. The bit rate then is that amount of time required to obtain one digital representation of the signal to the desired resolution. For example, moderately high-speed analog-to-digital converters can perform a 10-bit conversion in approximately 1 μsec (100 nsec/bit). The input signal (voltage) cannot vary any more than the LSB during the conversion period or the accuracy will be less than optimum. In this example the amplitude cannot vary more than $2^{-10} \times 100\%$, or from Table 7-6, approximately 0.1% during this period.

In a dynamic situation, the analog input signal is changing while the analog-to-digital conversion is taking place. The resulting digital value represents some analog value which existed during the conversion period.

Thus there is a limiting or maximum value in both frequency and amplitude that a sine wave can have which will still permit one to obtain a digital accuracy compatible with the change in amplitude of the sine wave during the conversion period. The time it takes for this compatible conversion is called the *aperture* time—in this particular example 1.0 μsec. It is a simple matter to show that for a sine wave of amplitude A and frequency f its maximum velocity is $V_{max} = 2\pi fA$. Hence if t_a is the aperture time the maximum permissible change in amplitude, ΔA, during t_a is

$$LSB = 2\pi ft_a \qquad (7\text{-}15)$$

since $\Delta A \leq LSB \times A$. If the signal amplitude is 10 volts and for our example, $t_a = 1$ μsec and $LSB = 0.000976$, (7-15) yields a maximum frequency of approximately 155 Hz.

A frequency of 155 Hz is too low for most practical dynamic applications. Higher sampling rates are achieved by using a sample-and-hold amplifier. This device can be considered a capacitor which stores the voltage to be measured so that the voltage level will not change (by more than the LSB) during the conversion period. With the sample-and-hold, such parameters as sample data entry, speed of operation of the sample-and-hold device, and accuracy requirements involving such items as the time required to change the hold capacitor from one value to another (settling time) and that of the capacitor after the storage has occurred (drift) have to be considered. For illustrative purposes and, in fact, even for practical applications, the sample-and-hold can be depicted as a capacitor to ground which has an input buffer amplifier with a switch which disconnects it from the capacitor and an output buffer amplifier to prevent stored voltage leakage. See Figure 7-19. In this particular case, the parameters involved are the settling time, which is the time for the capacitor to move from plus full-scale voltage to a minus full-scale voltage and settle out within the *n*-bit accuracy consideration; drift or leakage from the capacitor; and time for operation of the switch which puts the sample-and-hold into a hold mode.

From this viewpoint, aperture time is that time elapse between the signal given for the sample-and-hold to go into the hold mode until the complete transition has taken place. This operation takes place after the capacitor is tracking the data or, in other words, after the settling time has allowed a true

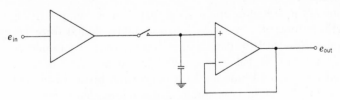

Figure 7-19 Sample-and-hold amplifier.

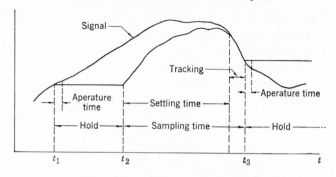

Figure 7-20 Data sampling using sample-and-hold.

"following" condition. Typical aperture times for sample-and-hold devices are of the order of 1 to 100 nsec. Thus for our example if we have a 10-nsec aperture time the highest frequency that can be considered and still be consistent with the quantizing error is 15,500 Hz.

Let us now examine the sample-and-hold operation in more detail. Prior to time t_1, as shown in Figure 7-20, the device is in the sample or follow mode. At time t_1, the sample-and-hold is put into the hold condition. After the sample-and-hold is put into the hold condition, it maintains this voltage within the appropriate accuracy limits until time t_2 by which time the conversion has taken place. The time t_2 is signified by an end of convert signal from the analog-to-digital converter and returns the sample-and-hold into a follow or sample mode. It now takes a maximum of $t_s = t_3 - t_2$, called the settling time, to be within the necessary follow specifications on a worst condition basis; namely, from plus full scale to minus full scale. At this time, the sample-and-hold is put into the hold mode and again conversion takes place. Typical settling times are of the order of 1 to 2 μsec for the following mode to be within 0.01% of the actual value.

In the present example a 10-bit word is obtained for each sample. Therefore, the cycle time (or throughput time) is that of 1-μsec conversion time plus the settling time, t_s, plus the aperture time. If $t_s = 1$ μsec the total time involved is then 2.01 μsec. This period is equivalent to a 500,000 10-bit word/sec conversion rate.

In summary it should be emphasized that with the sample-and-hold device the length of time of conversion into the actual digital form becomes unimportant with regard to quantizing accuracy, since the amplitude has been captured and can be held for a relatively long period of time, without significant loss in amplitude until the conversion process is complete. Also, the requirement that the maximum frequency be determined by (7-15) so that the signal does not change by ΔA during the conversion period was computed on a worst condition basis. Furthermore, this requirement is only to state what this frequency is in order that the results be consistent with the n-bit word length. If this condition is violated the quantizing error increases; that is, the effective quantization interval has been increased with a subsequent decrease in resolution. Finally, on the basis of Figure 7-4 the actual sampling rate should be on the order of 3 to 5 times the cutoff frequency of any band-limited signal in order to greatly reduce the aliasing error. In the above example if the sampling frequency were 3 times the cutoff frequency, the cutoff frequency is approximately 5150 Hz. It should be noted that the sample frequency can be greatly increased if the desired resolution is decreased (LSB increased). Considering the previous example, if we now require that a $n = 8$-bit conversion take place then without sample-and-hold the maximum frequency becomes 777 Hz since t_a is now equal to 0.8 μsec. If the sample-and-hold device is used the maximum frequency becomes 62,200 Hz. As an aside it should also be mentioned that the input impedance on the successive-approximation-type converters are low impedance, on the order of 2 to 10 kΩ. Thus the sample-and-hold amplifiers also act as a buffer (impedance) amplifier.

As a final remark it is noted that it is sometimes important to use a well-regulated power supply. Consider the case where there is a 0.05% change in the converter output for a 1% variation in supply voltage. In the case of a 12-bit converter, we see from Table 7-6 that this change in supply voltage can produce a 2-bit error in the converter output.

7-8   Digital-to-Analog Conversion (Decoding)

A typical digital-to-analog converter is shown in Figure 7-21. The operational amplifier is used in a summing mode (recall Section 3-11.4). Each input terminal is fed one bit of information at some voltage level V_0 and is connected to the amplifier through a voltage divider. The values of the

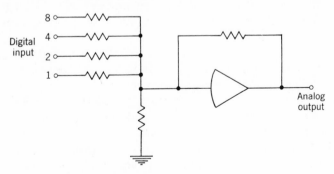

Figure 7-21   Digital-to-analog converter.

resistors in the voltage dividers are so selected that an input to the "8" terminal will produce twice as much output from the amplifier as would have been produced by the same amount of the input to the "4" terminal. Likewise, the resistor in series with the "4" input is so selected that an input here will produce twice as much output from the amplifier as would have been produced by an input to the "2" terminal and so on. As a result, the output of the amplifier is an analog voltage that corresponds to the digital input. If, for example, inputs are applied to the "1," "4," and "8" terminals (binary 1101), the output of the amplifier will be proportional to $8V_0 + 4V_0 + V_0 = 13V_0$.

Probably the most useful characteristic describing a digital-to-analog converter is its settling time. It is defined as the amount of time from when an input number is changed to when the output settles to within some specified accuracy of its final value. Settling time should include all the delays it is subjected to, including transients, switch delays, slew rate limitations, and bandwidth limitations. Also, in high-speed converters the switching transients should be kept to a minimum, on the order of less than the LSB.

7-9   Time-Division Multiplexing

In Section 7-2 we discussed the sampling of band-limited signals and found that a minimum sampling frequency f_s was necessary to obtain all the information required to faithfully reproduce the original signal. In the time domain this corresponds to taking a sample every $T_{sa} = 1/f_s$. If this sampled amplitude is converted into a digital number in t_c and furthermore, if the pulses corresponding to the binary number are transmitted serially from the output of the converter such that the total of the n pulses in the time axis is t_p, then if $T_{sa} \gg t_c$ and $T_{sa} \gg t_p$ we see that there is a large portion of the time, $T_{sa} - t_c - t_p$, that is available to transmit other messages. When many messages are propagated over a common medium by allocating different time intervals for the transmission of each signal the process is called time-sharing or time-division multiplexing.

Multiplexers are usually classified according to whether they are high-impedance or constant-impedance types. In a constant impedance multiplexer, as shown in Figure 7-22a, the inputs are connected, through resistors, either to signal ground (OFF) or to the virtual ground point of the operational amplifier (ON). Therefore, the input load on each source is constant, regardless of whether or not that particular source is selected to be digitized. This type of multiplex is useful for multiplexing inductive loads, scaling unequal input voltage ranges, and high-level input signal ranges (to 100 V).

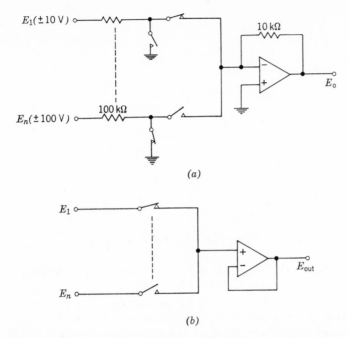

Figure 7-22 (*a*) Constant impedance and (*b*) high-impedance-type multiplexers.

In a single-level high-impedance multiplexer, all channels are switched directly to a common mode that is connected to the high-impedance input of a buffer amplifier, as shown in Figure 7-22*b*. When a channel is selected, the impedance of the switch that connects the source to the amplifier is relatively low, while the impedances of the nonselected switches are extremely high. Therefore, the buffer amplifier will impose an insignificant loading effect on the source. However, a dynamic error can be caused by connecting many multiplexer channels to a common point. A typical value of capacitance of an OFF channel is 7 pF (7×10^{-12} F). For an N-channel multiplexer, the total capacitance will thus be approximately $7N$ pF. If the impedance of each source connected to each multiplexer channel is R_s, the source resistance and the OFF capacitance form a low-pass filter with a time constant $\tau = 7NR_s \times 10^{-12}$ sec. If the source resistance is 10^4 ohm, for example, and $N = 64$ (that is, a 64-channel multiplexer) then $\tau = 4.5\ \mu$sec. Since approximately nine time constants (τ) are required for a step function to settle to within 0.01% of its final value, the settling time would be of the order of 40 μsec. This magnitude of settling time would greatly limit the system throughput capability.

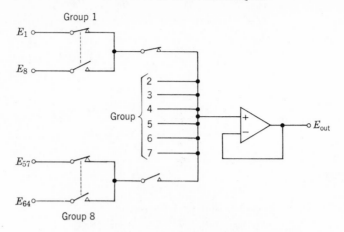

Figure 7-23   Two-level multiplexing.

To overcome this difficulty a two-level multiplexing technique can be advantageous. With this technique, the signal is routed through two multiplexers which connect it to the buffer amplifier. For example, 64 input channels might be accommodated using nine 8-channel multiplexers arranged as shown in Figure 7-23. Eight of the multiplexers comprise the first level. The outputs of these multiplexers are, in turn, connected to the buffer amplifier through the ninth 8-channel multiplexer which comprises the second level. In operation, one channel in each input group of eight in the first level is turned on. The second-level multiplexer then selects only one of the outputs from the first level. This technique provides the advantage that the total OFF capacitance is now the total of the seven OFF channels of the first level and the seven OFF channels of the second level. Hence the total capacitance has been reduced in the current example by a factor of $\frac{7}{32}$, and therefore, the settling time decreases by this amount with a corresponding increase in the throughput capabilities. Thus the settling time in the present example is reduced to 9 μsec.

Returning now to the introductory paragraph of this section we see that an additional time must be subtracted from the total available time left in which one can transmit other messages; namely the settling time, t_s. (Switching time is usually negligible compared to t_s.) Thus the number of different messages, M, that can be transmitted serially over the common channel is $M = T_{sa}/(t_c + t_p + t_s)$. If the transmission is parallel then $M = T_{sa}/(t_c + t_s)$.

In time-division multiplexing each message (pulse or group of pulses) occupies the entire bandwidth of the system and as a result all messages are transmitted with equal fidelity. It is this property that makes sampling techniques superior to other types of multiplexing methods. To deduce this

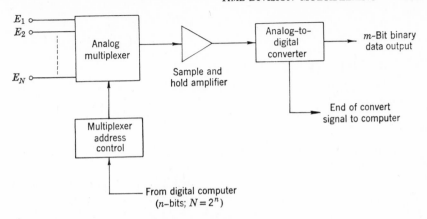

Figure 7-24 Multiplexer digitizer.

bandwidth the following is noted: the sampling theorem states that $2B$ samples/sec/message are needed or $2MB$ samples/sec, where again M is the number of messages to be transmitted along the common channel. From the study of low-pass filters in Chapter 2 it is recalled that when a step voltage is applied to a low-pass filter with cutoff frequency f_c, the peak of the response is reached at approximately $t = 1/2f_c$. Hence a recognizable pulse (with output amplitude proportional to input amplitude) results at the output if the cutoff frequency is the reciprocal of the pulse duration. Thus it is concluded that the transmission of samples by pulses requires a bandwidth of $2MB$ Hz for parallel transmission and $2nMB$ Hz for serial transmission (n is the number of bits).

Most multiplexers are used in conjunction with sample-and-hold amplifiers and analog-to-digital converters and furthermore are often controlled by a digital computer that determines which channels are selected and at what rate. This type of system is shown (without the computer, which is discussed in a subsequent section) in Figure 7-24. The N analog signals are connected to the input of the multiplexer. The particular channel to be digitized is sent in binary form from the computer to the multiplexer address control which then converts this information into a form which switches (gates) in the appropriate channel. The number of binary bits, n, required by the address control for N input channels is determined by $N \leq 2^n - 1$ wherein zero binary is used to denote a channel. Thus if $N = 64$, then $n = 6$ (0 to 63). The signal then goes to the sample-and-hold amplifier and then to the converter. The converter, upon completion of the conversion, gives an end of convert signal, which when sensed by the computer will then transmit the binary information to the computer input storage register. The cycle is then repeated for a new or the same analog input channel.

7-10 Counters

7-10.1 Measurement of Frequency

The basic principle by which counters are used for the precise determination of frequency is shown in Figure 7-25a. The input signal whose frequency is to be measured is converted into pulses by means of a Schmitt trigger (recall Section 3-11.6). These pulses are fed into one input of the main gate. To determine the frequency it is now only required to keep the gate open for a known interval of time. The clock for the timing gate interval is an accurate crystal oscillator whose frequency, f_c, is divided by 10^n in n discrete steps. The frequency of the pulses from the dividers is selected by externally choosing the time base which puts a state 1 on one of the inputs of one of the AND gates and leaves a state 0 on all the others. This signal is then fed into another Schmitt trigger whose output goes to the main gate. Thus the length of time the gate stays open is determined by which divided frequency output is selected.

Consider a sine wave of frequency f_0. The output of the Schmitt trigger is a series of pulses of frequency f_0 which are one input to the main gate. The other input is from the time base via the Schmitt trigger. The positive output of the Schmitt trigger is controlled by the frequency of the square wave, $f_c/10^n$, from the time base. Hence this input to the AND gate stays in the 1 state for a time duration $t_1 = 10^n/f_c$ and the total number of input pulses to the decade counter is $t_1 f_0 = 10^n f_0/f_c$. Since in most counters $f_c > f_0$, to obtain the most accurate frequency count n should be as large as possible. Notice that if $t_1 = 1.0$ sec and $f_0 = 10$ Hz the counter only records 10 pulses. On the other hand if $f_0 = 1000$ Hz, the counter records 1000 pulses. Since one source of error in a counter results from the fact that a variation of ± 1 count may be obtained, depending on the instant when the first and last pulses occur in relation to the sampling time, it is seen that the 10-Hz signal measured with a 1-sec time interval could give a 10% error at this frequency. Notice, however, that for the 1000-Hz signal the error due to the ± 1 count variation yields 0.1% error. Another source of error arises from jitter in the dividers, which may produce a small uncertainty in the sampling period. A third source of error is due to the signal itself. Since the Schmitt trigger is used to convert the signal to pulses the signal itself must have very little noise in it or the resulting pulses will be distributed in a random manner. In many counters a signal-to-noise ratio of 40 dB is required to keep this trigger error less than 0.3%. Beyond these, of course, the accuracy depends on the accuracy of the crystal oscillator.

With regard to crystal accuracy, there are two descriptions usually employed: long-term stability (also called crystal aging rate or drift) and

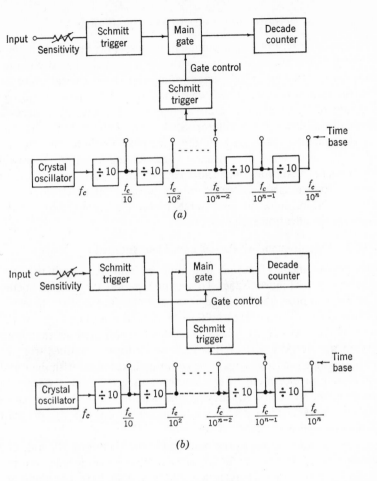

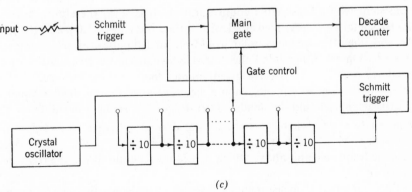

Figure 7-25 (*a*) Counter used to determine frequency, (*b*) time interval, and (*c*) time interval with period averaging.

short term stability. Long-term stability refers to slow, but predictable, variation in average oscillator frequency with time due to changes occurring in the crystal itself. After an initial period, aging in a good crystal becomes slow and assumes a predictable linear characteristic. The slope of this line is the aging rate of the oscillator. Since aging is cumulative, it is necessary to periodically calibrate the oscillator. Short-term stability indicates the effect of noise generated internally in the time-base oscillator on the average frequency over a short time, usually 1 sec. A long averaging time will hide large frequency variations. The short-term effects are so small that the short-term stability is a useful indicator only in those counters which contain very stable oscillators since in less stable oscillators, other errors make this short-term specification insignificant.

7-10.2 Measurement of Period and Time Interval

The measurement of period and time interval use essentially the same circuit shown in Figure 7-25b. To measure period the gate is opened and closed at some prescribed voltage level of the leading edge of the wave form and then closed by the same voltage level of the leading edge of the next period of the wave form. The voltage level is adjusted so that the count starts on the steepest part of the signal wave form to reduce trigger error. During the time the gate is open the counting takes place with the number of counts being proportional to the length of time the gate is opened. Again consider a sine wave of frequency f_0. The Schmitt trigger converts this signal into a square wave whose period is $t_1 = 1/f_0$. Thus one input to the main gate is a positive pulse (1 state) of duration t_1. During this time $f_c/10^n$ pulses appear at the other input to the main gate, and therefore, the output of the main gate is $t_1 f_c/10^n$ pulses which are counted in the decade counter. It is seen that for the greatest accuracy that n should have the smallest value consistent with a reasonable count duration. It is seen, therefore, that if a 10-Hz signal ($t_1 = 0.1$) measured with a time interval of 1.0 μsec ($10^n/f_c = 1.0$ sec), then the decade counter would record 10^5 pulses for an accuracy of 0.001%. On the other hand the 1000-Hz signal will be measured with an accuracy of 0.1%. Thus for a given counter there is some frequency below which period measurements should be used and above which frequency measurements should be made. That frequency is determined from the relation $f_0 = \sqrt{f_c}$, where f_c is the crystal oscillator frequency and f_0 is the frequency below which it is more accurate to use the period mode to determine frequency and above which it is more accurate to use the frequency mode.

For time-interval measurements between two signals the gate is opened by the first pulse in the same manner as described in the period measurement above and closed in the same manner by the second pulse. The count is again proportional to the time elapsed between both signals. This technique in effect measures the phase between two periodic signals. If the frequency of the signal is f, then the phase difference θ, in degrees, between the two signals is $\theta = 360 fT$, where T is the time delay between the two signals as measured by the counter.

An improvement to the period measurement which reduces error and improves resolution is multiple-period averaging. This is a very simple technique and is shown in Figure 7-25c. The input signal is divided by 10^n. This divided quantity controls the frequency of the state 1 of the Schmitt trigger output. The other input to the main gate is the undivided oscillator clock pulses. Depending on the number of periods (in this case, multiples of 10) that are averaged, the greater the resolution of the period measurement. In most counters the decimal point on the output display is automatically shifted to give the reading as if it were for a single period.

Application 7-1   Shear Modulus of a Linear, Viscoelastic Material

Viscoelastic materials are those materials in which the material constants (tensile, bulk, and shear modulus) are time (frequency) dependent. Many polymers fall into this class as opposed to common metals at room temperature which are elastic. In performing a material property test it is very desirable to choose a geometry and loading such that in describing the system mathematically only one material constant appears in the input-output (stress-strain) equation. One geometry and loading which isolates one material constant, the shear modulus, is the torsion of a cylindrical rod. If one applies a harmonically varying torque (moment) at one end and records the torque amplitude and phase response at the other end, the shear modulus as a function of frequency can be obtained over a narrow frequency band.

A block diagram of the experimental set-up is shown in Figure 7-26. The oscillator signal is fed to an amplifier whose output impedance is approximately equal to that of the vibration exciter [recall (3-18)]. The linear motion of the vibration exciter is converted into rotational motion by the lever arms attached to the top and bottom of the specimen. (This is a reasonable assumption since the magnitude of motion of the exciter is less than 100 μin.) The output from each accelerometer is fed into a switching box and then into an amplifier with high input impedance (recall Application 3-3). One output from the amplifier is fed into a variable bandpass filter and then into a narrow-band filter, amplifier, and detector. The other channel is just passed through the narrow-band filter, amplifier, and detector. The output from these second amplifiers are then fed into a counter set in the time interval mode and to an oscilloscope with a dual-trace unit.

Over the frequency range of interest the electronics behind the accelerometers introduce a relative phase shift between channels, which varies from

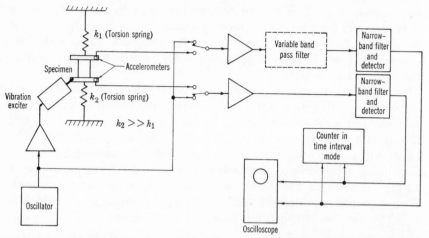

Figure 7-26 Functional diagram to obtain shear modulus of a linear, viscoelastic material.

frequency to frequency. This phase difference can be canceled by passing the same signal from the oscillator through both channels. With the same amplitude recorded on the meters of the narrow-band filters, which were tuned to the particular frequency, the variable bandpass filter's cutoff frequencies are adjusted until the time interval recorded on the counter is zero (or ± 1 digit) on its smallest scale (i.e., microseconds or nanoseconds). Without changing any of the setting of the electronics the accelerometer signals are switched in. The amplitudes from each channel are recorded and the phase determined from the time interval (recall Section 7-10.2). Whether the recorded phase angle is $+$ or $-$ can be determined from the oscilloscope trace or from the physics of the problem, e.g., minus before the first natural frequency. The procedure is repeated over the usable frequency range.

The variable bandpass filter need not be used. Instead the phase between the two channels is simply measured and recorded when the same signal is passed through each. When the accelerometer signals are switched in, the new time interval has added to it or subtracted from it the reading obtained previously. The only disadvantage with this technique is determining whether the previous reading should be added or subtracted.

7-11 Digital Voltmeters

7-11.1 Introduction

A digital voltmeter converts a slowly varying dc voltage of unknown magnitude into an observable decimal number. This displayed number is precise and discrete and does not have to be interpolated. Hence a digital voltmeter has an inherent accuracy and freedom from ambiguity. As we shall see from the subsequent discussion, the digital voltmeters also provide high-noise rejection, thus reducing the uncertainty of the reading.

The digital voltmeters can be used to record the value of rapidly varying signals by first converting the signal to a dc value using the detectors described in Chapter 4 (with long averaging times in the case of random signals) and then feeding this slowly varying dc value to the digital voltmeter.

7-11.2   Error Considerations

Since digital voltmeters have overall accuracy from 0.5 to   0.005% it is important to investigate the various possible sources of error.

1. Accuracy Statement.   Accuracy is usually specified in two parts: a percentage-of-reading error and a percentage-of-full-scale error. The former is a fixed error, independent of the input signal whereas the latter varies with the magnitude of the input signal. A part of this percentage-of-full-scale error is the quantizing error. From Table 7-6 we see that a three-digit digital voltmeter has a minimum fixed error of 0.05% while that with four digits has an error of 0.003%. Hence a typical digital-voltmeter accuracy statement would read "$\pm x\%$ of reading $\pm y\%$ of full scale."

Let us now examine two ways in which this accuracy statement can be interpreted. Let the percentage or fraction of the total range be α ($0 \leq \alpha \leq 1.0$). Then the total error ϵ_r as a function of α is $\epsilon_r = (y + \alpha x)\%$. Thus ϵ_r varies from $y\%$ to $(x + y)\%$ in a linear fashion. On the other hand, if we examine the ratio of the error ϵ_r to α we have $\epsilon_i = \epsilon_r/\alpha = (y/\alpha + x)\%$. Thus at full scale the error $\epsilon_i = \epsilon_r = (x + y)\%$; but at quarter scale we find $\epsilon_i = (4y + x)\%$. To illustrate this we consider the two statements: (1) $\pm 0.05\% R \pm 0.05\% FS$ and (2) $\pm 0.090\% R \pm 0.01\% FS$. At full scale the errors are equal. However, at one-tenth scale case (1) gives $\epsilon_i = 0.55\%$ whereas case (2) gives $\epsilon_i = 0.19\%$. Hence a comparison of accuracy statements is best made on a worst case basis. Obviously, in either case it is best to use the device as near to full scale as possible.

2. Input Impedance.   As was indicated in (3-13b) and Figure 3-11 it is important that the input impedance of the digital voltmeter be as high as possible so that a reasonably wide range of source resistances can be accommodated with very low measurement errors. Another source of error comes from the input amplifier offset current and voltage (recall Section 3-11). If i_0 is the offset current and Z_i is the source impedance, then the voltage error, v_0, due to the offset current is given by $v_0 = i_0 Z_i$. Thus if the offset current was 10^{-9} A and the source impedance was 10^6 ohms, then $v_0 = 1$ mV. If the digital voltmeter is being used in the 1-V range an error of 0.1% is introduced. However, if the digital voltmeter has an input impedance of 10^{10} ohms, the loading error due to the 10^6-ohm source would be only 0.01%. Thus under certain conditions offset current can cause significant errors.

3. Temperature Effects. Like the accuracy statement, the temperature-coefficient statement is usually of the form ($\pm$% of reading $\pm z$% of full scale)/°C. As in the case of the accuracy statement, when comparing these error coefficients it should be done on a worst case basis.

4. Noise. A digital voltmeter reading is subject to two types of noise error: common-mode and superimposed. As discussed in Section 3-11.5 common-mode voltages are caused when the digital voltmeter is grounded at a different potential than the source. Isolation from ground of the input circuit prevents these grounding problems. The amount of indifference to the grounding problems is specified in terms of common-mode rejection. A complete discussion of the various techniques used to alleviate this problem is given in Appendix C. Consider a digital voltmeter that has 120-dB common-mode rejection at 60 Hz. Then a 2-V peak-to-peak 60-Hz common-mode voltage would appear as a ± 1 digit on the 100-mV range of a five-digit meter. If the common-mode rejection was only 100 dB, there would be ± 2 digits of uncertainty.

Superimposed (or "normal-mode") noise is all other types of noise present in the input signal. This noise is eliminated by either filtering the signal or by integration. When integration is used the device is called an integrating digital voltmeter. This technique is discussed in the next section. When filtering is used and sampling speed is important, filter settling time should be known. As in the case of common-mode rejection, normal-mode noise rejection (sometimes abbreviated as NMNR or NMR) should be as large as possible.

7-11.3 Digital-Voltmeter Techniques

Digital voltmeters are classified according to how they convert the analog signal into a digital number. Discussion of four currently used techniques follows.

1. Successive Approximation. This technique was discussed in Section 7-7.2. The advantages of this type of digital voltmeter is high reading speed (100 readings/sec), fixed encode time regardless of the magnitude of the signal, and good long-term stability. The disadvantage of this technique is that it lacks inherent noise rejection which must be obtained by prefiltering the input signal.

2. Ramp. This technique was discussed in Section 7-7.1. The disadvantages of this technique are the inability to achieve a truly linear ramp and its lack of noise rejection, which must be obtained by prefiltering.

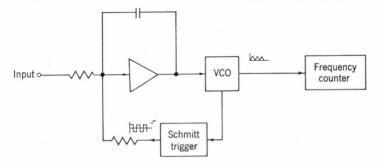

Figure 7-27   Voltage-to-frequency digital voltmeter.

3. Integrating: Voltage-to-Frequency Conversion.   Referring to Figure 7-27 it is seen that the input signal (almost dc) is converted to a train of sawtooth waves by means of a voltage-controlled oscillator whose frequency of oscillation is determined by the average level of the input signal. If the VCO is centered at $f_0 \gg 0$, then a negative voltage will be a frequency less than f_0 while a positive voltage will give a frequency greater than f_0. The number of pulses generated by the VCO during the fixed time-base period are totalled by the counter. Since the VCO, pulse generator, and integrator are in a closed-loop feedback system, the rate of pulse generation is such that the average value of the rectangular pulse train is equal to the dc input voltage. Furthermore, the accuracy, linearity, and stability of this type of conversion are largely determined by the width- and height-determining portion of the pulse generator and are independent of the stability and linearity of the VCO.

The advantage of this type of technique is that it tends to average out noise having a period small compared to the integration time and, in fact, this technique exhibits theoretically infinite rejection of frequencies whose periods are an integral multiple (or submultiple) of the integration time. A disadvantage in certain cases is that although the readout speed is moderately fast it does not equal that obtained by the successive approximation technique.

4. Integrating: Dual Slope.   A simplified diagram of a dual-slope digital voltmeter is shown in Figure 7-28. The input dc signal E_{in} is integrated for a known fixed time T_1. Thus, at the end of T_1 the voltage level is $E_1 = -E_{in}T_1/RC$. After the fixed time interval T_1, the unknown input is replaced by a precision reference voltage E_{ref} and the integrating capacitor discharges to zero at a known constant rate in time T_2. During the time T_2 the gate is opened and the counter records the number of pulses from the oscillator, $N_c(N_c = f_0 T_2)$. Since $E_1 = E_{ref}T_2/RC$ we find that $E_{in} = C_1 N_c$, where

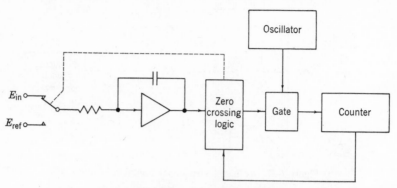

Figure 7-28 Dual-slope digital voltmeter.

$C_1 = E_{ref}/T_1 f_0$, a known constant for a given digital voltmeter. Hence the number of counts is directly proportional to E_{in}.

The advantage of the dual-slope technique is that it tends to cancel systematic errors during a measurement so that the components do not require as much precision and stability as the VCO technique. Because of the integrator it has the same noise rejection properties as the VCO technique.

7-12 Digital Delay Line/Transient Recorder

The shift register discussed in Section 7-6.3 can be combined with an analog-to-digital converter and a digital-to-analog converter to form a very useful device to act as either a digital delay line or a transient recorder as shown in Figure 7-29. The input signal is converted by the analog-to-digital converter to an n-bit binary number. The binary number is transferred in parallel to the input buffer register. Both the sampling instruction and

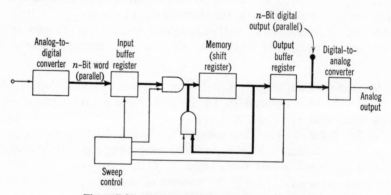

Figure 7-29 Digital delay line/transient recorder.

transfer to the input buffer are controlled by the sweep control unit. From the input buffer the binary word is entered (in parallel) into the M word memory (shift register). The output of the shift register goes to the output buffer register and an AND gate, controlled by the sweep control unit. The output of this AND gate goes to the input of the memory. The output from the output buffer register can be tapped off as a digital number or passed through a digital-to-analog converter which gives back the analog signal.

When used as a digital delay line the input signal is delayed in time by M/f_s, where f_s is the sampling frequency. In this mode the information in the memory is continually being updated. When used as a transient recorder the approximate duration of the pulse must be known. From Figure 1-7 we see that most of the frequency information in a pulse is below $f_c = 1/\tau$, where τ is the pulse duration. In order that the memory is fully utilized, however, $f_s = Mf_c$, where f_s is the sampling rate. Then the shortest duration pulse would be of the order of M/f_s. Using some appropriate level so that the input signal triggers the sweep control unit, the transient signal is "captured." After being stored in the memory the binary information can be recirculated indefinitely with the analog signal continually represented at the output of the digital-to-analog converter. Notice that this reproduced pulse could be "played back" at a different rate so that different readout devices can be used. Typical values of n and M are 6 and 100, respectively, with sampling rates up to 10 MHz.

7-13   Transient Peak Voltmeter

The peak-voltmeter circuit described in Section 4-2 is severely limited to signals with a high-duty cycle. For low-duty cycles and single transients of short duration (1 μsec) that type peak voltmeter is difficult to make work. An entirely different technique to record the peak value of low-duty cycle signals or transients is shown in Figure 7-30. A wide-band operational

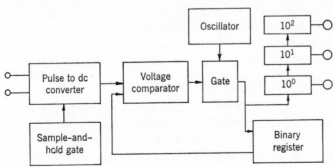

Figure 7-30 Digital peak voltmeter.

amplifier is used to charge the storage capacitor C. Since the capacitor is in a closed-loop amplifier circuit it may be charged at very high rates (the input rate) and the charging current "snapped-off" upon reaching the peak voltage. This is done by controlling the discharge time of the storage capacitor after it has reached its peak value (i.e., after the input voltage rate of amplitude change has reversed). In other words, the input amplifier is a modified sample-and-hold device which holds the peak. This peak value is held only long enough to convert the information (a short-term dc voltage) to digital form. This conversion is formed by applying this short-term dc voltage to a voltage comparator. The other input to the comparator is a binary-coded weighted feedback voltage from a 10-bit binary register. If the dc voltage is higher in amplitude than the feedback voltage, the comparator opens the gate allowing an oscillator to add to the binary register contents until the feedback voltage rises to the value of the dc level. At that point the comparator closes the gate. The number of counts is summed by the decade counter scaled to correspond to the amplitude. Since a 10-bit binary number has resolution to 0.1%, the decade counter is only three decimal digits. Furthermore, 1000 pulses are counted when the input signal is a maximum; hence the shortest time between input sample (or duty cycle in the case of periodic signals) is $\tau_m = 1000/f_0$, where f_0 is the oscillator frequency.

7-14 Small Word-Length Digital Computers

7-14.1 Introduction

In this section we describe the basic features and typical operations of small word-length (8-, 12-, 16-, and 18-bit) computers. These types of computers are used in numerous applications such as instrumentation, communication, and control systems. They are used as small stand-along engineering and scientific computers, but more frequently they are used in real-time applications. Several special-purpose computers designed to perform specific tasks such as spectrum analysis and digital data averaging are described in the next section. However, these special devices will essentially operate according to the techniques described in this section.

7-14.2 Computer Timing and Memory Cycles

Referring to Figure 7-31 we see that there is an internal timing generator which determines the speed of operation of the entire system. This timing generator automatically generates a read/write instruction (code) every $x \, \mu\text{sec}$. The term "read" denotes that information is retrieved from memory; "write" denotes storage of information in memory. The reading-out of a piece of information and then the writing-in of a piece of information

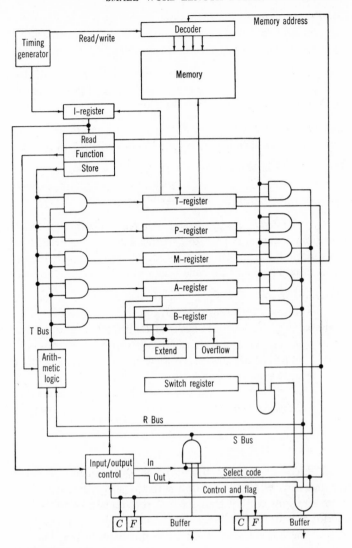

Figure 7-31 Simplified functional diagram of a typical small word-length computer.

constitutes a full cycle, sometimes termed a memory cycle. In this case x μsec is the memory cycle time of the computer and is one of the fundamental (although sometimes misleading) indicators of computer speed.

Most memory cycles have several operating modes. Given below are several types and their particular function.

1. Read/Restore Mode. Information stored at a particular location in memory is read out to the T register; then that same information is written back into the same memory location. This is a full-cycle operation.

2. Clear/Write Mode. Information stored in a particular location in memory is cleared (erased, by reading it without saving it), then new information is written into that memory location. This is a full-cycle operation.

3. Half-Cycle/Read Mode. During each half-cycle interval, information is read out from some memory location. No information is restored.

4. Half-Cycle/Write Mode. During each half-cycle operation, information is written into some selected memory location. The locations used will have been cleared of information in an earlier operation.

5. Split-Cycle (Read/Modify/Write) Mode. Information stored at a selected memory location is read out to the T register, where it is modified and the resultant information is restored, or written back into that same memory location. This is a full-cycle operation.

6. Read-Only Mode. This is a special case of mode 1, the read/restore mode. Whenever any information is read out, it is also automatically written back into the same address and is therefore continually available for reference. Thus memory (or a portion of it) is used only to provide readouts; the stored information is not modified, erased, replaced, or supplemented—hence the "read only" designation.

The modes described above are not mutually exclusive, and they may be employed in any desired sequence for a virtually unlimited variety of applications.

7-14.3 Working Registers

T-REGISTER. This holds the information transferred in or out of memory. Once a word of information is in the T-register, it is accessible for arithmetic operations and for transfers to other registers (via the bus system). For the write operation, the T-register is loaded by transfers from the other registers, and the information is stored in memory during the latter half of the memory cycle.

P-REGISTER. This holds the address of the instruction to be obtained from memory. It is also the computer's program counter which means that this register goes through a step-by-step counting sequence that causes the computer to read successive memory locations, corresponding to the existing count. This is generally a one or two memory-cycle operation. This register can also be made to skip a memory location or transfer to another point in memory altogether.

M-REGISTER. This holds the address of the location in memory to be accessed. This register is the only means of addressing specific memory locations. The setting on the M-register can occur from any of the other registers. Referring back to the P-register, we see that the computer really transfers the desired address from the P-register to the M-register, which in turn addresses the desired memory location. Thus it could be assumed that the P-register directly addresses memory, although in fact it goes through the procedure just described. Both the M- and P-registers are needed so that one register (P) can keep track of the current instruction in the case where the M-register may have to be changed several times in the course of executing an instruction.

A-REGISTER. This accumulates the results of arithmetic operations and serves as a transfer register for input and output. This register is called an accumulator.

B-REGISTER. This is a second accumulator, with same functions as A-register. The reason for having more than one accumulator is to provide faster, more flexible arithmetic than can be provided with only one accumulator.

EXTEND. This is a one-bit register used to indicate a carry from the A- or B-register or to link the A- and B-register. If the computer is an n-bit machine, then this latter function serves to form a single $2n$-bit word. The former function is seen by again considering a n-bit word. The maximum quantity stored in the register is n-ones. Then if one is added to this binary number the result would be n zeros which is obviously incorrect; it is correct if the EXTEND bit, which is now in the 1 state, is temporarily assumed to the bit n. The program can be written so that this assumption is valid. In order that this EXTEND bit be interpreted correctly, it should be set to the 0 state before each addition.

OVERFLOW. This is a one-bit register to indicate arithmetic overflow in the A- or B-register. The difference between the EXTEND register and this register is that the OVERFLOW register indicates that the largest "signed" quantity has been exceeded. Since bit n is the sign bit, bit $n - 1$ is the source of the significant carry. Having two possible signs means that detection of overflow requires two different sets of conditions. For addition of two positive numbers, overflow occurs if there is a carry from bit $n - 1$ to bit n in one of the accumulators. For addition of two negative numbers (which is in complement form), overflow occurs if there is not a carry from bit $n - 1$ to bit n.

In general, the more working registers a computer has the more flexible the programming can be and the faster information to and from the computer can be transferred.

7-14.4   Bus System

Referring to Figure 7-31, the computer shown is an "R-S-T" bus configuration. This is a conventional notation designating a three-bus system which applies to input buses (R and S) to an arithmetic unit with the output on the third bus (T). The use of two input buses permits arithmetic operations combining the contents of the two registers. Note that several register combinations are possible as inputs to the arithmetic unit. In some computers the A- and B-registers are addressable as memory locations, and therefore, the contents of these registers can be transferred via the R and T buses into the T-register. From this point, the contents of the T-register can be combined with either accumulator, including combining the number with itself. It should be realized that the R-S-T buses transmit the binary information in parallel.

7-14.5   Instruction Logic

As shown in Figure 7-31 the m-bit $(m < n)$ instruction register (I-register) receives its information from the T-register by reading the first m significant bits during the read/restore mode and decoding the instruction. The decoded instruction enables three functional operations described below, which in turn become active at specific times, depending on the instruction.

1. Read.   The read signal, connected to the output gate of all the five working registers, strobes the data of one or two registers onto their corresponding buses (R and S). This places the data at the input of the arithmetic logic circuits.

2. Function.   The function signal activates one of the arithmetic functions. The selected function alters or combines the data on the R and/or S buses and writes the resulting data out to the T bus.

3. Store.   The store signal, connected to all the input gates of the working registers, effectively opens the input gate of one or more of these registers to accept the data which appears on the T bus [from (1) above].

7-14.6   Word Length, Format, and Memory Addressing

Small digital computers are characterized by their instruction set (repertoire) and word length. In other than stand-alone-type scientific computer applications, the precision required in a specific application (or group of applications) determines the data word length. The instruction and length is a function of the number of instructions necessary to utilize fully the capabilities of the

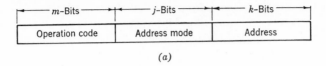

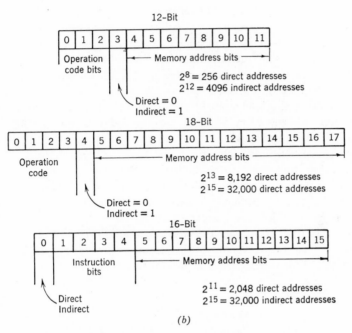

Figure 7-32 (a) Generalized instruction word format and (b) typical instruction word formats for 12-, 16-, and 18-bit computers.

computer. It is usually desirable to have the data word length and the instruction word length the same so that data and instructions can be stored interchangeably in memory.

In a computer, it is necessary to specify the required number of instruction types in an abbreviated code and to know how to address the full memory with an abbreviated address field. In Figure 7-32a appears a generalized instruction word format. If m bits are allocated for the operation code, then the computer is limited to a 2^m-instruction repertoire, unless some exceptions are taken to conventional instruction format concepts. One such exception is to set up two categories of instructions—those that address memory and those that do not. Then all but one of the available operation codes can be assigned to memory reference instructions which use the address field for addressing purposes. The remaining operation code can be used to represent the entire category of instructions that do not reference the memory (e.g., register-to-register transfers). This then permits use of the address field to specify the operation to be performed, that is, as an extension of the operation code.

For specifying address modification modes, such as direct and indirect, many computers provide one or more bits in the address-mode field to permit two or more address modes. The remaining bits make up the addressing field, which is obviously insufficient for directly addressing the full memory. Figure 7-32b shows typical bit allocation in 12-, 16-, and 18-bit computers.

Direct addressing of memory implies the use of a single memory-instruction location. In the examples shown in Figure 7-32b it should be mentioned that only one half of the directly addressable memory locations are available. The remaining direct memory locations are the "page 0" locations (defined in section 7-14.8), a set of fixed locations at one end of the computer memory. In indirect addressing, the memory location specified by the memory reference instruction contains the address of the memory location actually desired. An instruction using indirect addressing requires an additional memory cycle to obtain the address and it requires an additional memory location. The major difference then between direct and indirect memory addressing is that the larger the number of directly addressable memory locations, the faster the speed of operation of the computer. This difference, however, exists only if the particular program uses a large amount of indirect addressing to arrive at far-afield memory locations.

As we have implied a computer operates by going through a list of instructions (stored program) that are sequentially decoded. The decoding of these instructions informs the hardware logic of the computer to remove information from specific memory locations, manipulate the information, and return the information to memory. In some instances this latter step is altered to send the information instead to some input-output device. The memory reference instructions, therefore, describe how to manipulate this information and sequence it through the stored program. These memory reference instructions typically take two memory cycles. From Figure 7-32b we see that there are eight possible memory-reference instructions with a 12-bit word and sixteen possible memory-reference instructions with either 16- or 18-bit words. Because of its large repertoire of memory-reference instructions, 16- and 18-bit computers should require fewer memory locations to write out a sequence of programming steps. In many applications, however, the advantages of superior software (e.g., FORTRAN and BASIC computer languages) outweigh the speed loss and memory loss due to the small word length.

The set of instructions that do not reference memory are called micro-instructions, register reference instructions, register-to-register instructions, and the like. These are used to test data in the accumulator and make logical decisions about what operations the computer should do next. Also, these instructions shift information from one accumulator to another. Such instructions are probably used most frequently in real-time applications. It is possible to make computers with any number of microinstructions since it is a matter of computer hardware and has nothing to do with computer word length. Typically there are between 40–50 microinstructions in the 12-, 16-, and 18-bit machines.

7-14.7 Computer Arithmetic Operation

In small computer arithmetic operations there are three basic types of numbers: single-precision, double-precision, and floating point. A single-precision number is determined by the word length of the computer. Thus the largest number available in single-precision is 2^{n-1}, where n is the number of bits. The reason for the exponent of $n - 1$ is that one bit is reserved for the sign. Hence with a 12-bit word the largest number is, from Table 7-6, ± 2048; 16-bit word yields $\pm 32,768$; and 18-bit word yields $\pm 131,000$. Thus the single-precision 12-bit word is a 3-digit (decimal) number with 100% overrange; the 16-bit word is a 4-digit number with 300% overrange; and the 18-bit word is a 5-digit number with 30% overrange. If two words of memory (core) are used with every arithmetic operation the largest number is 2^{2n-1} and the resulting number is called a double-precision number.

A major factor affecting single- and double-precision arithmetic operations in a real-time computer system is the lack of a decimal point. It is possible to handle decimal numbers with an operation known as scaling; however, it requires large memory allocations and substantially longer computer operation times. If one does not require a decimal point the word length is determined by the best resolution of the data entering the computer. Thus as was done to obtain the maximum digital error in Table 7-6, the percentage resolution is given by $100\%/(2 \times 10^m)$, where m is the number of the decimal digit accuracy. Thus, for example, considering a 12-bit single-precision word length the percentage accuracy of the input need not be more accurate than 0.05%.

In order to overcome the inability of single- and double-precision numbers to handle easily decimal points a floating-point number system is used. A floating point-number system allows one to specify a number of decimal digits multiplied by 10 to a power. Referring to Figure 7-33 we see that several possible ways exist in which floating-point numbers can be coded and their corresponding digits. We see that the difference between the 12-, 16-, or 18-bit word lengths is not in the size of the floating-point number system but rather in the difference in the accuracy between four, six, or eight digits. A two-word floating-point system is probably 30 to 40% faster than a

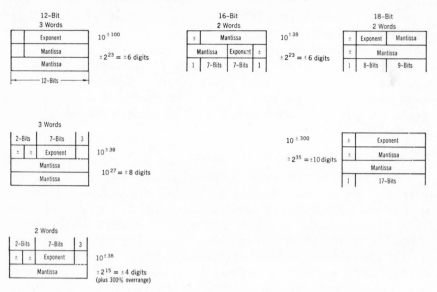

Figure 7-33   Floating-point word formats.

three-word floating-point system. Therefore, with six-digit information some advantage accrues with a 16-bit word-length computer. In terms of memory required, it obviously requires 50% more word-memory location to store data with a three-word floating-point system. In some systems the accuracy of the floating-point-number system can easily be changed with the change in accuracy requirements.

7-14.8   Computer Memory

The memory stack is a fundamental component, in which data are stored, and from which they are retrieved. In itself, it is incapable of performing any function except that of holding the data; it must be operated upon by active, programmed elements that can cause it to accept and relinquish data according to a plan, that can time and sequence all of the individual steps required, and that can perform all of the conversions, translations, interpretations, and adaptations necessary to enable full, compatible communication between the external equipment and the stack. This combination of stack and essential interface constitutes the memory system or "memory."

One feature of a computer memory is its "access mode." The access mode describes the sequence in which its individual data-storage addresses may be programmed automatically, in any series of operations. The ability of the computer to operate in more than one access mode multiplies the versatility of the memory enormously. There are three types of access modes which are described as follows.

1. Random Access.   In this mode, address locations are selected through an internal register that can be set to any location in the memory stack; addresses can be selected in any desired sequence.

2. Sequential Access.   In this mode, the memory is addressed by the same address register used for random access, but under the control of an internal address counter that advances the register by one count after each memory cycle (recall the function of the P-register). All memory locations are thus addressed in sequence. It should be noted that the sequential addressing can begin and end wherever desired.

3. Sequential Interlace Access.   In this mode, the address register is controlled by two independent address counters, thus giving independent control of the read and write functions of the memory cycle. Hence information may be written in any series of addresses and interlaced with reading out other previously stored information at any other series of memory addresses. It should be apparent that the random-access mode is inherent in all memories; sequential and sequential interlace modes are provided by adding the address counter(s) necessary to program the address register.

Another term used to describe memory systems is "page" size. Page size refers to a fixed number of contiguous memory locations which can be addressed by the address field. The "page" and the addresses in it are relative to some reference (i.e., program counter, fixed memory location, index register). Typical page sizes range from 256 to 2048 words. Two common techniques for page addressing are page relative addressing and addressing relative to the program counter. The former method is a page-addressing concept where the page boundaries are fixed and the page used is usually determined by the higher-order bits of the program counter. In the latter technique the page extends a half page size before and after the current address in the program counter.

Core memory stacks are constructed in a variety of configurations (or "formats"), according to the number of drive lines employed in writing-in a data bit at a memory address. By far the most widely used is the 3D 4-wire configuration, which works in the following manner. A strong current pulse in a wire passing through a magnetic ferrite core causes the core to be magnetized in a certain direction; this constitutes storage of a 1 at that address bit. A strong pulse in the opposite direction causes the magnetic polarity to reverse; this stores a 0 at that location. A weak pulse (half-strength or less) does not affect the magnetic polarity of the core.

Pulses are used also in reading out the information stored. A pulse in a given direction will cause a core that has been magnetized in its 1 direction to reverse its polarity, thus reverting to the 0 state; in doing so, it will induce a small pulse in a separate ("sense") wire that is also threaded through the

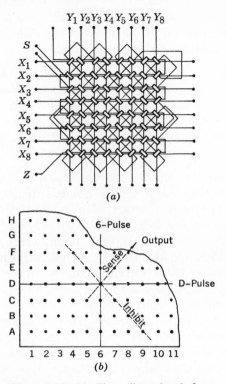

(a)

(b)

Figure 7-34 (a) Three-dimensional four-wire plane with sense and inhibit winding, and (b) simplified memory plane.

core; this output pulse indicates that a 1 was stored at that location. If the core was already in the 0 state, the read pulse will have no effect on the magnetic polarity; no pulse will be induced in the sense line; this absence of an output pulse indicates that a 0 was stored at that location.

We now explore the procedure by which a 3D 4-wire stack utilizes this function in storing and retrieving digital data. Figure 7-34a shows a wiring diagram of a 64-core plane in this format; note that each core (the cores are shown edgeways) is threaded by four wires. The S ("sense") wire passes through all cores; the Z ("inhibit") wire also passes through all cores; each core is threaded by an X-wire and a Y-wire—note, however, that no two cores share the same X-Y combination. Figure 7-34b represents a portion of a memory plane, wired as shown in Figure 7-34a, but simplified to permit examination of the operating principle. Viewed from above, the memory stack consists of an orderly assembly of data-storage addresses, arrayed as in

Figure 7-34b, which represent only a small segment of the stack; here we have identified the rows by letters, and the columns by numbers, for convenience in discussion. Each dot represents a location where one bit of data—either a 1 or a 0—can be stored. Assume that all addresses initially have 0's stored in them.

To store a bit at any selected location or address requires sending two current pulses through this location simultaneously, via the only two co-ordinate drive paths that cross at this location. To store a bit at address D6, as shown in the figure, requires a pulse through row D, and a pulse through column 6. Individually, the pulses are not strong enough to affect any addresses that they pass through, but at the one point at which they coincide, address D6 in the example, their combined magnetic field is sufficient to effect a change in the magnetic field of the ferrite core at that location. This constitutes storage of a 1 at that location. If, instead, we wish to store a 0 at that location, we would send a pulse through another path (labelled INHIBIT in the Figure) that passes through all of the addresses, simul-taneously with the "D" and "6" write pulses. This inhibit pulse is not, in itself, strong enough to affect any address that it passes through; however, at D6 it acts to neutralize one of the write pulses. The residual field is thus not strong enough to influence the magnetic state of the core at this address, and the core remains in its 0 state. In either case, the application of the coincident currents has caused one bit of information to be stored at address D6. To store one binary-coded decimal numeral requires storage of four such bits; accordingly we would require three more layers, or planes, identical to the one shown in Figure 7-34b in order to store the numeral. The D-rows and 6-columns in all four planes would be pulsed simultaneously; to store a 0 at D6 in any one plane we would also pulse its inhibit line; to store a 1 we merely omit the inhibit pulse for that plane. Obviously, the plane can be extended in either or both coordinates to increase the word-storage capacity; bit length of the words is increased by adding planes.

To obtain data stored at a particular address in the memory, a procedure very similar to the above is employed. Signals are applied to the appropriate column and row as before, but in the form that will cause the address-bit to assume a 0 state. If the bit was previously in the 1 state, it will revert to the 0 state; in doing so it will induce a pulse signal on the sense line that passes through all addresses in the plane. If the bit was already in the 0 state it would remain so, and no pulse would be induced. When a data word is read out of a memory address, all bits at that address are left in the 0 state. To restore the original data, a 1 is written into each bit that previously contained a 1. The other bits simply remain in the 0 state.

Referring again to Figure 7-31 we see from the above discussion that each bit position of the T-register is wired by the sense and inhibit lines through all cores on the corresponding bit plane. Since only one core on an individual plane is sensed (addressed) at a given instant of time, the sense line needs only to detect a change in magnetic polarity anywhere on the bit plane. Similarly, the inhibit signal is applied to the entire bit plane when writing, but actually affects only the selected core.

7-14.9   Software

The most important software for small computers is the assembler. The assembler generates the binary machine code. This machine code can be absolute or relocatable. Absolute code refers to specific memory locations, whereas relocatable code allows a program to be loaded into any convenient set of available memory locations. A relocatable assembler is one that can generate relocatable code. A routine commonly called a relocatable loader assigns the available absolute memory addresses to the program prior to execution. The loader loads the program into memory, allocates memory space for all necessary subroutines associated with the main program, and links the subroutines together so they can be properly executed in the main program. In small memory computers (4096 words) where only one dedicated program is to be continually run, a relocatable feature is not a necessity, but in larger systems (8192 words and up) where many programs are run a relocatable assembler is very useful.

Both one-pass and two-pass assemblers are used. A pass refers to the assembler scanning the source program. A one-pass assembler does not require the source program to be reloaded, thus saving time, but it has some storage allocation limitations. A two-pass assembler allocates storage and sets up proper address reference label tables on the first pass. On the record pass is generated the basic machine code. It should be noted that the number of passes refers to the number of times the assembler scans the source program which is not necessarily the number of times the program is read in from the peripheral device. One convenient feature in many assemblers is the provision for pseudo-operation codes. This makes it easier to reserve memory locations for constants. The pseudo code is not a machine instruction but an instruction to the assembler itself. Compiler languages (e.g., FORTRAN, BASIC) are used infrequently in real-time applications of small-word-length computers because the input/output subroutine must be closely tailored to the specific characteristics of the external device when interfacing to external hardware. Tight timing requirements and higher machine efficiency also frequently dictate the use of assembly languages.

7-14.10   Input/Output

Information is transferred into the computer from an external device or out of the computer to an external device by way of its input/output system.

A transfer of information is initiated by a signal from a device indicating that it is ready for input or output. The transfer occurs by the process of interrupting a running program (of any type). The interrupt directs the computer to a location in memory uniquely associated with the interrupting device. This location, in turn, directs the computer to a program routine (previously stored in memory), and this routine will contain instructions which effect the actual transfer of information. Since interrupts can occur at almost any time, including during the routine of an earlier interrupt, a priority network is usually present in the computer to establish the sequence in which interrupts are serviced.

Almost all external devices have to be interfaced with the computer. This is usually done with what is known as interface cards. A typical interface card contains an input/output buffer consisting of n flip-flops (n corresponding to the computer word length) for the temporary storage of data to be transferred in or out, so that it is not necessary to tie up a working register during the relatively long transfer periods. The actual number of buffer bits will depend on the device for which the interface is intended. If the buffer is less than n bits, the data are transferred to or from the least significant bits of the A or B registers. The interface card also contains an input/output flag flip-flop, which will be set up by a signal from the external device when the device has completed an operation. The flag, in many systems, may also be set by a program instruction. Once the flag is set, it remains set until reset by a clear instruction from the computer. Provided it is itself not inhibited by the set flag or a higher priority device (or otherwise disabled), the flag, when set, inhibits all interrupts for devices having lower priority. It will cause an interrupt after the current memory cycle. Some systems offer successive interrupts for one device which may occur on receipt of a number of flag signals without executing a clear instruction, thus making it possible to inhibit lower priority devices indefinitely until a desired number of high-priority transfers have been completed.

The third important feature of an interface card is a control flip-flop to command or to enable the external device to perform its input or output operation. In addition, the control bit controls the interrupt capability for that particular device; that is, unless the control flip-flop is set, a received flag cannot cause an interrupt nor can it inhibit the interrupt capability of any other device in the priority string. Thus the control bit when set effectively "turns on" the individual input/output channel.

On computer command one or more external devices begin their read or record operations, putting data into (input) or taking data from (output) the input/output buffer on each individual interface card. During this time the computer may continue running a program or may be programmed to wait for a specific device. On completion of the read or record operations, each device returns an "operation completed" signal (flag) to the computer. The flags are passed through a priority network which allows only one device to be serviced regardless of the number of flags simultaneously present. The flag with the highest priority causes an interrupt at the end of the current memory cycle. When this interrupt occurs, the computer puts the select code number of the interrupting device into the M-register, thus causing the next instruction to be read from the memory location having the same number as the select code. This location in memory is referred to as the "interrupt location" and is reserved for that particular device. At the same time, the read/restore mode is set so that the computer must execute the instruction contained in the specified memory cell. Generally this instruction will be a jump to a service routine. This subroutine consists of instructions that will prepare or accept the new data. On completion of service, it is the subroutine's responsibility to return the P- and M-registers to the values they contained before being interrupted.

Let us now return to Figure 7-31. The interface arrangements are shown for only two external devices, one input and one output. The switch register is shown as part of the input/output system and is considered to be an input device. As indicated in the figure, the input/output control logic is used to process all input/output operations in two ways: process input/output instructions and process service requests by peripheral devices. These two operations are discussed separately in what follows.

The input/output instructions are decoded by the instruction register and routed to input/output control, which translates the instruction into appropriate driving signals. One such signal is the "in" signal which strobes all interface positions for input (represented by two AND gates in Figure 7-31, one accepting data from a buffer register and one accepting data from the Switch-register). Only one of these interface positions can be enabled according to the select code in the T-register, and the corresponding data are strobed by the "in" pulse onto the S bus. From there it is transferred via the T bus into the A- or B-register (as enabled by a store signal at the A or B input gate). Another driving signal is the "out" signal. This signal strobes all interface positions for output. The select code from the T-register enables one interface position and permits the "out" signal to strobe the data on the R bus into the corresponding output buffer. (The data on the R bus were read out of the A- or B-registers by a read signal.)

In addition to transferring data the input/output control can (according to programmed instructions) send out signals to test the state of the control and flag bits (C and F) or to set or reset these bits. The select code determines which interface will receive the signal from the input/output control. The control and flag bits are command signals for transferring data between the buffer and the peripheral device.

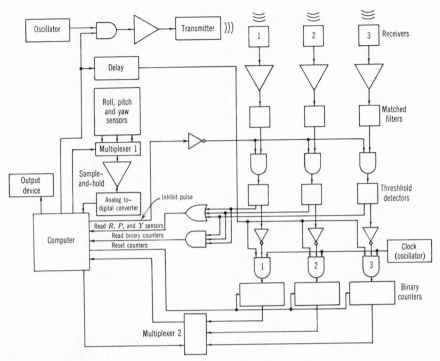

Figure 7-35 Functional diagram of a short-base navigation system.

Application 7-2 Short-Base Navigation System

A short-base navigation system is a system whereby one can determine the location of a submerged object relative to the surface ship with three-hull-mounted acoustic receivers (hydrophones). The receivers are mounted on the ship's hull at the furthest distances feasible from each other. One of the receivers acts as a transmitter. The transmitter transmits a sinusoid of duration τ_0 and frequency f_0. The submerged object is equipped with a device (called a transponder) which receives this signal and after several milliseconds transmits a signal of duration τ_1 and frequency f_1. From the time of arrival of the pulse to the first receiver its hits, the relative times of arrival between the other two receivers and the first, and the orientation of the ship with respect to the earth coordinates, the location of the submerged object with respect to the ship can be found. If the geographic coordinates of the ship are known then those of the submerged vehicle can easily be determined.

A block diagram of a short-base navigation system is shown in Figure 7-35. The computer is used as a controller and interpreter of the information. At predetermined intervals the computer gates the oscillator which feeds the transmitting hydrophone (shown here as a separate device). Since it takes a finite time for the short-duration sinusoid to build up to maximum value, this same gate pulse is delayed by this amount of time before it reaches the AND gates 1 to 3. When it arrives at these gates the binary counters start their count. At the same time an inhibit pulse prevents the electronics behind the three hydrophone receivers from responding to the transmitted pulse. After a preselected time period the inhibit pulse is removed.

The submerged object receives the transmitted pulse and responds with a return acoustic signal. This signal is received by one of the hydrophones, amplified, and passed through a matched filter to optimize the peak signal-to-noise ratio. The output is then passed through a threshold detector set at some level, which may be adjusted depending on the magnitude of the incoming signal. The output of each threshold detector stops its respective counter. At the same time, the output of the first threshold detector to receive a signal sends a signal to the computer which instructs it to read the ship's orientation sensors: roll, pitch, and yaw. These signals are multiplexed through a sample and hold amplifier to an analog-to-digital converter. After the signal has arrived at all these hydrophones, a pulse is transmitted to the computer to read the binary counters. After the counters have been read a reset pulse sets them to zero.

Having all the necessary data, the computer determines the relative location of the object with respect to the surface ship. The information is then displayed in some convenient fashion. After a preselected period of time the computer initiates the process and new data are obtained.

7-15   Signal Averagers

Two techniques have been already discussed which can be used to increase the signal-to-noise ratio in periodic signals: correlation and filtering. In the latter method, if the repetitive signal has a large bandwidth, the filter is virtually useless in improving the signal-to-noise ratio. A third technique is signal averaging. Signal averaging is a technique that takes advantage of redundant information provided by repetitive signals.

Signal averagers sample the input signal at fixed time intervals (see Figure 7-36), convert the samples to digital form, and store the sample values at separate locations in memory. Since we are sampling, the sampling rate must be at least greater than twice the highest frequency (with significant amplitude information) as described in Section 7-2. From Section 7-3 we see that it is desirable to filter out all frequencies beyond the highest frequency of interest. The sampling process is continued for a preset number of repetitions of the desired signal. During the first repetition, sample values are stored in memory, with each memory location corresponding to a definite sample time.

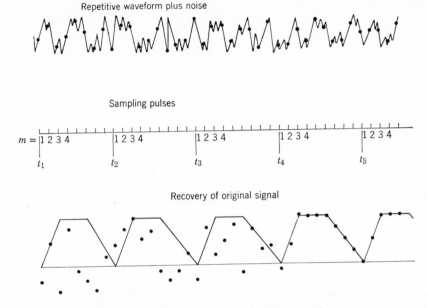

Figure 7-36 Signal averaging of repetitive wave forms.

Then, during subsequent repetitions, the new sample values are added algebraically to the values accumulated at the corresponding memory locations. After any given number of repetitions, the sum stored in each memory location is equal to the number of repetitions times the average of the samples taken at that point on the desired input wave form.

This simple summation process tends to enhance the signal with respect to the noise. The signal portion of the input is a constant for any sample point, so its contribution to the stored sum is multiplied by the number of repetitions. The noise, which is not time-locked to the signal, makes both positive and negative contributions at any sample point during the successive repetitions. Therefore the noise portion of the stored sum grows more slowly than the signal portion.

We now ascertain the signal-to-noise improvement mathematically. Let the input signal be $f(t)$ which is composed of a repetitive portion $s(t)$ and a noise portion $n(t)$. Then $f(t) = s(t) + n(t)$. Furthermore, we let the kth repetition of $s(t)$ begin at time $t_k(t_1 = 0)$ and the samples be taken every T sec. At an arbitrary sample point in time we have

$$f(t_k + mT) = s(t_k + mT) + n(t_k + mT)$$
$$= s(mT) + n(t_k + mT)$$

For a given k and m, $n(t_k + mT)$ is a random variable. We now assume that $n(t_k + mT)$ has a zero mean and an rms value of σ, and for different k, the noise samples are statistically independent.

Now consider the mth sample point. The peak signal-to-rms-noise ratio of any particular repetition is [recall (2-75)]

$$d = \frac{s(mT)}{\sigma} \tag{7-16}$$

After K repetitions, the value stored in the mth memory location is

$$\sum_{k=1}^{K} f(t_k + mT) = Ks(mT) + \sum_{k=1}^{K} n(t_k + mT) \tag{7-17}$$

Since the noise is random and the K samples are independent, the rms value of the sum of K noise samples is $\sigma\sqrt{K}$. Therefore, the signal-to-noise ratio after summation is

$$d_K = \frac{Ks(mT)}{\sigma\sqrt{K}} = \sqrt{K}\, d \tag{7-18}$$

where d is given by (7-16). Thus summing K repetitions improves the signal-to noise ratio by a factor of $\sqrt{K}$.

Returning to (7-17) we define

$$M_K{}^m = \frac{1}{K} \sum_{k=1}^{K} f(t_k + mT) \tag{7-19}$$

Then the average value stored in the mth memory location after K repetitions should be

$$M_K{}^m = M_{K-1}^m + [f(t_K + mT) - M_{K-1}^m]K^{-1} \tag{7-20}$$

Equation 7-20 is a relatively difficult algorithm to implement because of the large amount of logic circuitry required to perform a fast division by K. Furthermore, it is possible that round-off errors could build up to be a significant problem. A way around this problem is to approximate (7-20) by

$$M_K{}^m = M_{K-1}^m + \frac{f(t_K + mT) - M_{K-1}^m}{2^N} \tag{7-21}$$

where $2^{N-1} < K < 2^N + 1$. This equation is easy to implement because dividing by a binary number by 2^N is done by simply shifting its binary point N places to the left. Furthermore, round-off errors are eliminated by this algorithm. We see that (7-20) and (7-21) both give the same value for the signal portion of the average. However, a difference exists between the overall signal-to-noise ratio which increases to a maximum value of 0.77 dB as the number of repetitions increases. In other words, (7-21) gives a slight reduction of 0.77 dB in the overall signal-to-noise ratio as compared to (7-20) for a large number of repetitions.

If we were to examine the signal averager in the frequency domain, we would find that it behaves like a "comb" filter. That is, around every sample point there is a narrow-band filter whose bandwidth is approximately (see ref. 12 of Chapter 1, p. 318)

$$B \approx \frac{0.866}{K T_0} \text{ Hz} \tag{7-22}$$

provided that K is large. In (7-22) T_0 is the period or the repetition rate of the input signal.

Let us now illustrate these results with several examples. Since the signal averager is a digital device working in a binary system the number of repetitions is usually presented as $K = 2^n$ where $0 \leqq n \leqq 19$ is a typical range. From (7-18) we see that the signal-to-noise improvement d_K/d in decibels is $3n$ dB. Thus for $n = 19$ the increase in signal-to-noise ratio would be 57 dB. If the repetition rate is 500 Hz the bandwidth of the comb filter using (7-22) is $B \approx 0.000826$ Hz. Thus each contribution to the repetitive time wave form separated in time by the sampling interval T has been effectively filtered by a 0.000826-Hz bandwidth filter.

As a final remark it is pointed out that signal averaging of periodic signals is really not a new technique, but simply a practical implementation of cross-correlating a periodic signal with a periodic pulse train. In this regard recall Section 1-3.3 and see ref. 12 of Chapter 1, p. 303 ff.

7-16   Real-Time Spectrum Analyzers

7-16.1   Introduction

Numerous techniques exist for performing real-time spectrum analysis. On first sight the most straightforward method would appear to be to convert the analog signal into digital form and then feed the digital information to a digital computer which is programmed to take the Fourier transform of the signal. The programming technique is usually the fast Fourier transform* (FFT) or some variation thereof. This technique will not be discussed here. Instead we will concern ourselves with hybrid—partly analog, partly digital— analysis systems. These systems perform the actual spectral analysis with analog filters, but manipulate and store various intermediate and final values in a digital fashion. In the following sections two specific types of real-time analyzers are discussed; namely, $\frac{1}{3}$-octave and constant bandwidth.

---

* See refs. 6, 7, 8, and 11.

7-16.2   The ⅓-Octave Analyzer

The ⅓-octave filter discussed in Section 2-4.2 can be structured to perform as a real-time analyzer as shown in Figure 7-37. The input signal is simultaneously fed into the N parallel ⅓-octave filters. The output of each filter is fed into an rms detector, the output of which is integrated for a duration $T_k(k = 1, 2, \ldots, N)$. The voltage levels from each integrator is then multiplexed to either an analog-to-digital converter and processed further digitally, or it is kept in analog form and displayed on a recorder or large oscilloscope display unit.

From (2-38) the rise time of the bandpass filter, τ_r, is proportional to the bandwidth B. For a ⅓-octave filter, (2-58) gives that $B = 0.23 f_c$ where f_c is the center frequency. The lowest center frequency is typically 25 Hz and the highest 20,000 Hz. Thus the bandwidth varies from nominally 6 to 4600 Hz, and therefore τ_r varies from approximately 166 msec to 217 μsec. Furthermore, the normalized standard error, using (1-150), varies from $41.6/\sqrt{T_k}\%$ to $1.48/\sqrt{T_n}\%$, where T_k and T_n are the individual averaging times of the respective integrators. Since the longest rise time is 166 msec, this is the shortest possible time in which a sample can be taken. Furthermore, in order to ensure the same statistical accuracy regardless of the ⅓-octave band, the averaging time for $f_c = 25$ Hz must be 28 times longer than that at $f_c = 20,000$ Hz. Thus the specified statistical accuracy determines the longest integration time and hence the shortest possible meaningful sampling time of the low frequency filters.

A modification of this analog detection technique is to perform the detection digitally as shown in Figure 7-38. The output of each filter is continually scanned via the multiplexer such that each filter's output is sampled every Δt sec. Each sampled output from one scan of the multiplexer

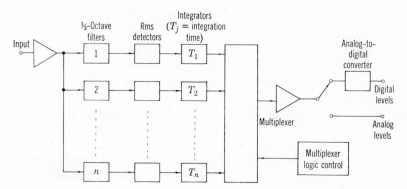

Figure 7-37   Real-time ⅓-octave analyzer using analog detection.

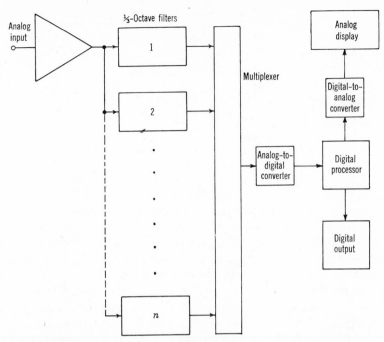

Figure 7-38   Real-time $\frac{1}{3}$-octave analyzer using digital detection.

is digitized, squared, and stored in one of the N memory locations. On the next scan the process is repeated with each new squared value being added to the previously squared value corresponding to the particular channel. The total number of samples taken, M, depends on the averaging time selected, T. The rms value then in each filter band is simply the square root of the sum of the squares in each corresponding memory location. This digital value of the sum of the squares can be converted to decibels and displayed in numeral form as a permanent record or it can be converted to analog form for display on an oscilloscope or X-Y recorder.

7-16.3   Constant Bandwidth Spectrum Analyzer

Consider the constant bandwidth spectrum analyzer shown in Figure 7-39. The input signal is passed through a low-pass filter and sampled at a rate higher than the minimum rate such that Figure 7-4 would give a negligible alaising and interpolation error. If f_s is the sampling frequency and f_c is the cutoff frequency, then from Figure 7-4, $f_s/f_c = 3$ and $k = 10$ would give less than a 1 % error. The output of the track-and-hold amplifier goes to an analog-to-digital converter where it is converted to an n-bit (including sign) binary number. The conversion time t_c is such that $t_c \ll 1/f_s$.

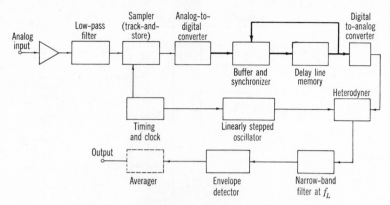

Figure 7-39   Real-time constant bandwidth analyzer.

The digital samples are then processed through a buffer storage and synchronization network into n parallel glass delay lines which function as the memory. The synchronized buffer assures that the data signals will appear in memory in the same sequence as the original signal. However, the recirculation of the memory is done in such a way that the adjacent samples following circulation are placed in the memory in a much closer proximity than the original sampling interval ($1/f_s$). Thus as data are circulated to the memory, they exit the delay lines at a much faster rate than they entered from the analog-to-digital converter. The signal has, therefore, been time compressed. If the total delay of the delay line is T_d and data can be clocked through these lines at a rate of f_r, then each line can store up to $N = f_r T_d$ bits with each bit exiting the delay line every $1/f_r$ sec. The effective speed up due to the time compression is $f_r/f_s = S_u$. Notice that it takes N/f_s long for the delay line memory to "fill-up." Each new data sample thereafter causes the first sample (taken N/f_s previously) to be discarded. Typical values for T_d and f_r are 100 μsec and 15 MHz, respectively. Thus if the $f_s = 10,000$ Hz, $S_u = 1500$. Note that with slower sampling rates the effective speedup is increased since the output rate of the delay lines is constant. It should also be mentioned that the buffer storages are used to store temporarily the sampled data so that $1/f_s$ can be less than T_d. This permits a small increase in the "real-time" high-frequency limit since f_s is usually three times the highest frequency of interest. When this limit is exceeded the device no longer works in real time since $T_d > 1/f_s$.

The output of the delay line memory is passed through a digital-to-analog converter which performs the entire conversion of the N n-bit words in T_d. The output of the converter is passed through a fixed low-pass filter to remove all sampling discontinuities. The analog form of the time compressed signal is now applied to a heterodyner. The other input to the heterodyner is from a very stable linearly stepped oscillator which varies in M programmed steps from f_L to f_H in $\Delta f = (f_H - f_L)/M$ steps. From (2-62) and the discussion in Sections 2-4.4 and 2-4.5 we see that a single "narrow-"band filter centered at f_L of bandwidth Δf will act as n bandpass filters provided that $\Delta f \approx 1/T_d$. We recall that T_d is the time required for the filter to respond to an input signal. Since the oscillator is stepped every complete time compressed cycle of T_d, the filter must be given at least T_d to respond fully. Typical values of M vary from 200 to 500, while those for f_L are 18 MHz. If the preceding value $T_d = 100$ μsec is used, $\Delta f = 10,000$ Hz, and if $M = 500$, $f_H = 23$ MHz. It should be noted that in order to convert correctly the output of the filter to power spectral density the effective bandwidth B_e must be used. Normally $B_e = c\,\Delta f$ where c is a constant greater than 1. Since the same filter is being used each time, c can be determined and the output of the filter suitably corrected when the division by B_e (attenuation) is performed.

The output of the bandpass filter is passed through an envelope detector. It can be shown* that if the output of the envelope detector is squared, the results for both a sinewave input and a random noise input are equal to that obtained by a true rms detector except that both are a factor of 2 too high. Thus this factor of 2 can be calibrated out by suitable attenuation of the output signal. A disadvantage of the squaring procedure is that its dynamic range is limited. To overcome this a pseudo-rms detection scheme is often used in conjunction with a signal averager. In this case the signal averager is used to average the output of the M filters. When summing statistically independent samples of the detector output, the output of the average possesses $2K$ degrees of freedom (K is the total number of samples averaged). The average from each filter is then squared (actually passed through a log converter and suitably scaled by a factor of 2) to give the pseudo-rms value. The only disadvantage of this technique is that it causes a 1-dB response differential between those signals which are sinusoidal in character and those which are random. Notice that the use of a signal averager permits one to work in effectively real time and still obtain results with a reasonable normalized standard error. From (1-151) we find that $n_e = 2$ since the BT product of each filter is approximately 1. Thus the averager obtains two degrees of freedom for each sample fed into it. Using (1-150) we find the $\epsilon = 100/\sqrt{K}\%$.

7-17 Summary

The two main parts involved in converting an analog signal to digital form are sampling and quantizing. In the case of an ideally band-limited signal the sampling principle states that if the time-varying function is sampled at

* See ref. 1.

twice its cutoff frequency the samples contain the complete and unambiguous information of the original signal. In the case of a bandpass signal the sampling frequency in the ideal case varies between two to four times the bandwidth of the original bandpass signal. The recovery of the original wave form from the sampled information can be obtained by passing the sampling pulses through an ideal low-pass filter for the case of a band-limited signal and through an ideal bandpass filter for the case of the bandpass signal. Using a sampling frequency that is less than the minimum stated above introduces aliasing errors which are caused by the downward spectral transposition of information. In most practical systems it is necessary to sample at rates higher than the minimum in order that the aliasing errors are minimized.

Quantizing is the representation of a continuous signal by certain discrete allowed levels. It inherently introduces an initial error which can be minimized by taking the quantization intervals small enough. Once a signal has been quantized it can be converted into a digital number. The most convenient form for the representation as a digital number, with respect to actual circuit construction, is the binary system. There are several forms of binary notation, the most common of which are straight binary, binary-coded-decimal (BCD), and the complement codes. The actual quantization of a signal is performed by an analog-to-digital converter which uses logic elements such as AND, OR, and NOT gates and flip-flops to convert a given amplitude into a binary number. The analog-to-digital converter often uses a sample-and-hold amplifier which captures the amplitude of a signal in nanoseconds and then holds it without significant loss of amplitude until the actual conversion is completed. The conversion scheme can be either a linear-ramp type to trigger a timing circuit in which the amount of time the gate is open is proportional to the amplitude, a successive-approximation type which compares the "captured" amplitude to known amplitudes, or a cyclic converter in which the decision in each step in the conversion process is independent of the result of the preceding step. After processing, the binary number can be converted back into an analog signal by a digital-to-analog converter. The digital-to-analog converter is simply a voltage divider network which gives the binary states the appropriate relative voltage weights which are then summed with an operational amplifier.

When the signal has been converted into digital form it is amenable to time-shared transmission over a common medium using time-division multiplexing. Multiplexers are usually classified as high impedance or constant impedance. In a constant-impedance type the input load to each source is constant, whereas in the high-impedance type all channels are switched directly to a high-impedance buffer amplifier. This latter method introduces a dynamic error due to capacitance effects which form a low-pass filter that greatly limits the system throughput capability. This difficulty is overcome somewhat by using a two-level multiplexing technique.

References

1. H. J. Bickel, "Calibrated Frequency-Domain Measurements," Monograph No. 2, Federal Scientific Corporation, New York, October 1969.
2. H. S. Black, *Modulation Theory*, D. Van Nostrand Co., Inc., New York, 1953.
3. R. B. Blackman and J. W. Tukey, *The Measurement of Power Spectra*, Dover Publications, Inc., New York, 1958.
4. E. Bukstein, *Digital Counters and Computers*, Holt, Rinehart and Winston, Inc., New York, 1960.
5. Y. Chu, *Digital Computer Design Fundamentals*, McGraw-Hill Book Company, Inc., New York, 1962.
6. W. J. Cochran et al., "What is the Fast Fourier Transform?," *Proc. IEEE*, **55**, No. 10, October 1967, pp. 1664–1674. Also *IEEE Trans. Audio and Electroacoustics*, **AU-15**, No. 2, June 1967, pp. 45–55.
7. J. W. Cooley, P. A. W. Lewis, and P. D. Welch, "Historical Notes on the Fast Fourier Transform," *Proc. IEEE*, **55**, No. 10, October 1967, pp. 1675–1677.
8. J. W. Cooley and J. W. Tuckey, "An Algorithm for the Machine Calculation of Complex Fourier Series," *Math. Comput.*, **19**, April 1965, pp. 297–301.
9. J. V. DiRocco, "Signal Conditioning for Analog-to-Digital Conversion in Instrumentation Systems," Electronic Instrument Digest, May 1970, pp. 7–11.
10. J. J. Downing, *Modulation Systems and Noise*, Prentice-Hall, Inc., N.J., 1964.
11. T. D. Enochson and R. K. Otnes, *Programming and Analysis for Digital Time Series Data*, U.S. Navy Publication and Printing Service Office, Washington Navy Yard, Washington, D.C. 20390, 1968.
12. J. C. Hancock, *The Principles of Communication Theory*, McGraw-Hill Book Company, Inc., New York, 1961.
13. J. L. Hughes, *Digital Computer Lab. Workbook*, Digital Equipment Corporation, Maynard, Mass., 1968.
14. H. V. Malmstadt and C. G. Enke, *Digital Electronics for Scientists*, W. A. Benjamin, Inc., New York, 1969.
15. J. Millman and H. Taub, *Pulse, Digital, and Switching Waveforms*, McGraw-Hill Book Company, Inc., New York, 1965.
16. P. F. Panter, *Modulation, Noise, and Spectral Analysis*, McGraw-Hill Book Company, Inc., New York, 1965.

17. R. K. Richards, *Electronic Digital Components and Circuits*, D. Van Nostrand Company, Inc., Princeton, N.J., 1967.
18. A. K. Susskind (Ed.), *Analog-Digital Conversion Techniques*, M.I.T Press and John Wiley & Sons, Inc., New York, 1957.
19. C. R. Trimble, "What is Signal Averaging?", Hewlett-Packard Journal, Hewlett-Packard Company, Palo Alto, Calif., April, 1968.
20. B. A. Weitz, "DVM Accuracy: Usable and Rated," Electronic Instrument Digest, September 1969, pp. 8–12.
21. "Using a Reversible Counter," Application Note 85, Hewlett-Packard Company, Palo Alto, Calif., 1966.
22. *A Pocket Guide to Hewlett-Packard Computers*, Hewlett-Packard Company, Palo Alto, Calif., August 1968.

GLOSSARY

The purpose of this glossary is to promote better communication through the use of a common concept of terms. It was compiled and assimilated from numerous sources, none of which claims the prestige and authority of a national or international committee on standards. It is, therefore, offered only as a convenience and not as dogma.

access, random. Access to computer storage under conditions in which the next position from which information is to be obtained is in no way dependent on the previous one.

access time. The time interval required to communicate with the memory or storage unit of a digital computer. The time interval between the instant at which the arithmetic unit calls for information from the memory and the instant at which the information is delivered. Also, the time interval between the instant at which the arithmetic unit starts to transmit information to the memory and the instant at which the storage of information in the memory is completed.

accumulator. The register and associated equipment in the arithmetic unit in which there are formed sums and other arithmetical and logical results; a unit in a digital computer where numbers are totaled, that is, accumulated.

accuracy. Freedom from error. The capability of an instrument to follow the true value of a given phenomenon. Accuracy contrasts with precision; for example, a four-place table, correctly computed, is accurate; a six-place table containing an error is more precise, but not accurate.

active element. Gain element.

active network. A network containing passive and active (gain) elements.

active RC network. A network formed by resistors, capacitors, and active elements.

address. A set of characters that identifies either a register, a location in storage, or a device in which information is stored; a label, usually in the form of numerical coordinates.

address, absolute. The label(s) assigned by the machine designer to a particular storage location; specific address.

alternate operation. A means of displaying on an oscilloscope the output signals of two or more channels by switching the channels, in sequence, after each sweep.

amplifier, buffer.   An amplifier used to isolate the output of any device from the effects produced by changes in voltage or loading in subsequent circuits.

amplitude.   The magnitude of variation in a changing quantity from its zero value. The word is usually modified with an adjective such as "peak," "rms," "maximum," and so on, which designates the specific amplitude in question.

amplitude response.   The maximum output amplitude obtainable at various points over the frequency range of an instrument operating under rated conditions.

analog.   An adjective which has come to mean continuous, cursive, or having an infinite number of connected points. The instrumentation industry uses the words "analog" and "digital" where the more precise language would be "continuous" and "discrete."

aperiodic signals.   All deterministic signals which are not periodic.

astigmatism.   In the viewing plane of the cathode-ray tube, any deviation of the indicating spot from a circular shape.

asynchronous inputs.   Those terminals in a flip-flop that can affect the output state of the flip-flop independent of the clock. Called set, preset, reset or clear.

attentuation.   Reduction or division of signal amplitude while retaining the characteristic wave form; implies deliberately throwing away or discarding a part of the signal energy for the sake of reduced amplitude.

azimuth loss.   A high-frequency loss in tape systems caused by misalignment of the reproduce gaps to the recorded magnetic energy. The effect is similar to widening the reproduce gaps.

balanced bridge.   A Wheatstone bridge circuit, which when in a quiescent state has an output voltage of zero.

balanced input.   A symmetrical input circuit having equal impedance from both input terminals to ground.

balanced line.   A transmission line consisting of two identical signal conductors having equal resistance and equal capacities with respect to the cable shield or with respect to ground.

bandpass filter.   A filter that transmits alternating current above and below its lower and upper cutoff frequencies, respectively, and substantially attenuates all other currents.

bandwidth.   A statement of the frequencies defining the upper and lower limits of a frequency spectrum where the amplitude response of an amplifier to a sinusoidal wave form becomes 0.707 (-3 dB) the amplitude of a reference frequency.

binary.   A characteristic or property involving a selection, choice, or condition in which there are but two alternatives.

binary-coded decimal (BCD).   A binary numbering system for coding decimal numbers in groups of 4 bits. The binary value of these 4-bit groups ranges from 0000 to 1001 and codes the decimal digits 0 through 9.

binary digit.   A digit in the binary scale of notation. This digit may be only 0 (zero) or 1 (one).

binary number.   A number written in binary notation.

binary-to-decimal conversion. Converting a number written in binary notation to one written in decimal notation.

bipolar. Having two poles, polarities, or directions. Applied to amplifiers or power supplies, it means that the output may vary in either polarity from zero.

bistable element. Another name for flip-flop. A circuit in which the output has two stable states (output levels 0 or 1) and can be caused to go to either of these states by input signals, but remains in that state permanently after the input signals are removed. This differentiates the bistable element from a gate also having two output states but which requires the retention of the input signals to stay in a given state. The characteristic of two stable states also differentiates it from a monostable element which keeps returning to a specific state and from an astable element which keeps changing from one state to the other.

bit. A binary digit; a smallest unit of information; a single pulse position in a group of possible pulse positions.

buffer. A circuit element which is used to isolate between stages or to convert input and output circuits for signal level compatibility. Storage between the input/output equipment and the computer where information is assembled in easily absorbed units; storage between the main memory and the computer where information is rapidly accessible.

bus. A path over which information is transferred, from any of several sources to any of several destinations.

calibration. The process of comparing a set of discrete magnitudes or the characteristic curve of a continuously varying magnitude with another set of curve previously established as a standard. Deviation between indicated values and their corresponding standard values constitutes the correction (or calibration curve) for inferring true magnitude from indicated magnitude thereafter.

capstan. The rotating shaft on a magnetic tape recording and/or reading device which is used to impart uniform motion to the magnetic tape on command.

cathode-ray tube. A large electronic vacuum tube with a screen for visual plot or display of output in graphic form by means of a proportionally deflected beam of electrons. A large electronic vacuum tube containing a screen on which information, expressed in pulses in a beam of electrons from the cathode, is stored by means of the presence or absence of spots bearing electrostatic charges.

chopper. An electromechanical or electronic device used to interrupt a dc or low-frequency ac signal at regular intervals to permit amplification of the signal by an ac amplifier.

chopped mode. A time-sharing method of displaying output signals of two or more channels with a single cathode-ray tube gun, in sequence, at a rate not referenced to the sweep.

chopping rate. The rate at which channel switching occurs in chopped mode.

chopper-stabilized amplifier. An amplifier configuration utilizing a carrier-type dc amplifier to reduce the effects of input offset and drift of a direct coupled amplifier.

closed loop. Implies overall feedback.

closed-loop bandwidth. The frequency at which the closed-loop gain drops 3 dB from its midband (or dc) value.

closed-loop input impedance. The impedance looking into the input port of an amplifier with feedback.

closed-loop output impedance. The impedance looking into the output port of an operational amplifier with feedback.

closed-loop voltage gain. The voltage gain of an amplifier with feedback.

code. The machine-language representation of a character. The instruction code is the set of symbols which conveys to the computer the operation which it is to perform. The instruction code always specifies a process; it usually specifies one or more operand addresses; it may specify the address of the next order; it may specify additional information such as a cycle index or breakpoint. The coded instruction code or machine-language operation code may sometimes be referred to as a code.

common-mode characteristics. The characteristics pertaining to performance when both inverting and noninverting inputs have a common signal.

common-mode input capacitance. The equivalent capacitance of both inverting and noninverting inputs with respect to ground.

common-mode input impedance. The open-loop input impedance of both inverting and noninverting inputs with respect to ground.

common-mode voltage range. The range of voltage that may be applied to both inputs without saturating the input stage.

common-mode rejection (CMR). The ratio of the common-mode input voltage to the differential input voltage which produces the same output signal. A measure of how well a differential amplifier ignores a signal which appears simultaneously and in-phase at both input terminals.

common-mode signal. A signal that appears simultaneously at both amplifier input terminals with respect to a common point.

comparator. A circuit which compares two stored codes and supplies an indication of agreement or disagreement; or a mechanism by means of which two items of information may be compared in certain respects, and a signal given depending on whether they are equal or unequal. A device for comparing two different transcriptions of the same information to verify agreement or determine disagreement.

comparison bridge. A type of voltage comparison circuit whose configuration and principle of operation resemble a four-arm electrical bridge. The elements are so arranged that a balance exists in the circuit and a virtual zero error signal is derived. Any tendency for the output to change in relation to the reference voltage creates a corresponding error signal, which, by means of negative feedback, is used to correct the output in the direction toward restoring bridge balance.

compiler. A program-making routine, which produces a specific program for a particular problem by the following process: (1) determining the intended meaning of an element of information expressed in pseudo code; (2) selecting or generating the required subroutine; (3) transforming the subroutine into specific coding for the specific problem, assigning specific memory registers, and so on, and entering it as an element of the problem program; (4) maintaining a record of the subroutines used and their position in the problem program; and (5) continuing to the next element of information in pseudo code.

complement. To reverse the state of a storage device or of a control level (e.g., changing a flip-flop from the 1 to the 0 state or vice versa). The complement of a number (base or base minus 1).

complementary tracking. A system of interconnection of two regulated supplies in which one (the master) is operated to control the other (the slave). The slave supply voltage is made equal (or proportional) to the master supply.

computer, asynchronous. A calculating device in which the performance of any operation starts as a result of a signal that the previous operation has been completed; contrasted with synchronous computer.

computer, synchronous. A calculating device in which the performance of all operations is controlled with equally spaced signals from a master clock.

constant current power supply (current regulator). A power supply capable of maintaining a preset current through a variable load resistance.

constant voltage power supply (voltage regulator). A power supply that is capable of maintaining a preset voltage across a variable load resistance.

core, magnetic. A magnetic material capable of assuming and remaining at one of two or more conditions of magnetization, thus capable of providing storage, gating, or switching functions, usually of toroidal shape and pulsed or polarized by electric currents carried on wire wound around the material.

counter. A device capable of changing states in a specified sequence upon receiving appropriate input signals. The output of the counter indicates the number of pulses which have been applied.

counter, binary. An interconnection of flip-flops having a single input so arranged as to enable binary counting. Each time a pulse appears at the input, the counter changes state and tabulates the number of input pulses for readout in binary form. It has 2^n possible counts, where n is the number of flip-flops.

crest factor. Ratio of the peak value of a signal to its rms value.

critical damping. The value of damping which provides the most rapid step function response without overshoot.

cross-talk. Interference in a given transmitting or recording channel which has its origin in another channel. Cross talk is normally not a constant value for all frequencies, being most severe at lower frequencies.

crowbar. An electrical short circuit intended to forcibly clamp the output voltage of a power supply.

current compensation.   A means of compensating the inevitable shunt conductance in circuits connected across the terminals of a current regulator, similar to the error-sensing provisions of a voltage regulator.

current limiting.   An overload protective mechanism which is intended for protection rather than constant current operation.

damping.   A characteristic built into electrical circuits and mechanical systems to prevent rapid or excessive corrections which may lead to instability or oscillatory conditions.

dc amplifier.   One which has a frequency response band that goes down to dc. This definition includes direct-coupled amplifiers but is not restricted to them.

decade.   A group or assembly of ten units.

decibel.   Abbreviated dB. Ten times the logarithm of the ratio between two amounts of power P_1 and P_2 existing at two points or at two instants in time. By definition,

$$\text{number of dB} = 10 \log_{10}\frac{P_1}{P_2} = 20 \log_{10}\frac{E_2}{E_1} + 10 \log_{10}\frac{Z_1}{Z_2}$$

assuming the power factors of the two impedances are equal. If the impedances themselves are equal, the right-hand term becomes zero and

$$\text{number of dB} = 20 \log_{10}\frac{E_2}{E_1}$$

deflection factor.   The ratio of the input signal amplitude to the resultant displacement of the indicating spot on an oscilloscope.

delayed sweep.   A sweep that has been delayed either by a predetermined period or by a period determined by an additional independent variable.

demodulator.   A device which extracts the modulation information from a modulated carrier.

density, packing.   The number of units of useful information contained within a given linear dimension, usually expressed in units per inch.

detector.   Converts an ac voltage into a slowly varying dc voltage that is proportional in some manner to the input voltage.

differential amplifier.   An amplifier whose output signal is proportional to the algebraic difference between two input signals.

differential flutter.   The difference in flutter measured between tracks on the tape caused by dynamic skew.

differential input.   An input circuit that rejects voltages which are the same at both input terminals and amplifies the voltage difference between the two input terminals. May be either balanced or floating and may also be guarded.

differential input impedance.   The impedance between the inverting and noninverting input terminals.

differential input voltage rating.   The maximum allowable signal that may be applied between the inverting and noninverting inputs without causing damage to the amplifier.

differential transducer. A device which is capable of following simultaneously the voltages across or from two separate signal sources and providing a final output proportional to the difference between the two signals.

digital output. An output quantity consisting of a set of discrete magnitudes coded to represent digits in a system of notation.

digitize. To render an analog measurement into digital form.

direct-record process. Data frequencies are linearly mixed with high-frequency bias. Features the widest bandwidth of magnetic tape recording systems.

discriminator. A device in which the properties of a signal such as frequency or phase are converted into amplitude variations.

distortion. An unwanted change in wave form. Principal forms of distortion are inherent nonlinearity of the device, nonuniform response at different frequencies, and lack of constant proportionality between phase-shift and frequency. (A wanted or intentional change might be identical, but it would be called modulation).

drift. A gradual and unintentional deviation of a given property. Variations which occur at a rate of 0.2 Hz or slower.

dual-beam oscilloscope. An oscilloscope in which the cathode-ray tube produces two separate electron beams that may be individually or jointly controlled.

dual trace. A mode of operation in which a single beam in a cathode-ray tube is shared by two signal channels.

dynamic range. A ratio, usually expressed in decibels, of the maximum signal with a specified distortion level to the minimum resolvable signal.

dynamic skew. The apparent variation in skew of the tape as it passes across the head, either in record or reproduce mode or both. See I.T.D.E.

electromagnetic. Pertaining to the mutually perpendicular electric and magnetic fields associated with the movement of electrons through conductors.

electrostatic coupling. Coupled by means of capacitance so that charges on one circuit influence another circuit owing to the capacitance between the two.

ensemble average. A statistical average performed over the members of an ensemble for one instant of time rather than over time for one signal. For an ergodic process the two are equal.

ergodic process. A stationary process where the time averages performed on one signal are equally representative of any other signal in the ensemble and are equal to the corresponding ensemble average.

error. The loss of precision in a quantity; the difference between an accurate quantity and its calculated approximation.

equivalent input wide-band noise voltage. The output noise voltage with the input shorted, divided by the dc voltage gain of the amplifier. (This voltage is measured with a true rms voltmeter and is limited to the combined bandwidth of the amplifier and meter.)

equivalent input offset voltage. The amount of voltage required at the input to bring the output to zero. Usually this voltage is adjustable to zero by using either a built-in or external variable resistor (balance control).

fall time. A measure of the time required for the output voltage of a circuit to change from a high-voltage level to a low-voltage level once a level change has started. Current could also be used as the reference; that is, from a high current to a low current level.

feedback. The returning of a fraction of the output of a machine, system, or process to the input, to which the fraction is added or subtracted. If increase of input is associated with increase of output, subtracting the returned fraction (negative feedback) results in self-correction or control of the process, while adding it (positive feedback) results in a runaway or out-of-control process.

feedback elements. The elements used to feed a portion of the output signal back to the input (also called feedback network).

feedback attentuation. An attenuation factor in the feedback loop by which the output voltage is attenuated to produce the input error voltage.

fetch. The portion of a computer cycle during which the location of the forthcoming instruction is determined, the instruction obtained and modified if necessary, and the instruction entered into the control register.

flip-flops (storage elements). A circuit having two stable states and the capability of changing from one state to another with the application of a control signal and remaining in that state after removal of signals. (see bistable element.)

flip-flop, D. D stands for delay. A flip-flop whose output is a function of the input that appeared one pulse earlier; for example, if a 1 appeared at the input, the output after the next clock pulse will be a 1.

flip-flop, JK. A flip-flop having two inputs designated J and K. At the application of a clock pulse, a 1 on the J input and a 0 on the K input will set the flip-flop to the 1 state; a 1 on the K input and a 0 on the J input will reset it to the 0 state; 1's simultaneously on both inputs will cause it to change state regardless of the previous state. $J = 0$ and $K = 0$ will prevent change.

flip-flop, RS. A flip-flop consisting of two cross-coupled NAND gates having two inputs designated R and S. A 1 on the S input and 0 on the R input will reset (clear) the flip-flop to the 0 state, and 1 on the R input and 0 on the S input will set it to the 1. It is assumed that 0's will never appear simultaneously at both inputs. If both inputs have 1's it will stay as it was; 1 is considered nonactivating. A similar circuit can be formed with NOR gates.

flip-flop, RST. A flip-flop having three inputs, R, S, and T. This unit works as the RS-flip-flop except that the T input is used to cause the flip-flop to change states.

floating. The condition of a device or circuit that is not grounded and not tied to any established potential.

floating input. An isolated input circuit not connected to ground at any point. It is understood that in a floating input circuit both conductors are equally free from any reference potential, a qualification which limits the types of signal-sources which can be operated floating.

flutter. An instantaneous tape-speed variation caused by mechanical effects such as torsion, stiction, and eccentric rollers. It is a measure of the dynamic speed errors produced in transporting the tape expressed in peak-to-peak error within a bandpass for each speed. Wow is an audio flutter of the low-frequency variety and is not normally used in instrumentation terminology.

FM recording process. Data are used to modulate a carrier frequency and are applied to the tape near saturation. Features the best overall data quality with least distortion, widest dynamic range, effective freedom from amplitude instability, ease of time base change, and phase linearity.

frequency-modulated (FM) signal. A signal where the information is contained in the deviation from a center frequency. In a recording instrument, this deviation is proportional to the applied stimulus.

frequency response. The portion of the frequency spectrum which can be covered by a device within specified limits of amplitude error.

full power frequency. The maximum frequency at which the amplifier can produce a full-rated output voltage swing within a specified total harmonic distortion.

full scale. The total interval over which an instrument is intended to operate. Also the output from a transducer when the maximum rated stimulus is applied to the input.

gate. A circuit which has the ability to produce an output which is dependent upon a logical function of the input.

gate, AND. All inputs must have 1 level signals at the input to produce a 1 level output.

gate, NAND. All inputs must have 1 level signals at the input to produce a 0 level output.

gate, NOR. Any one input or more than one input having a 1 level signal will produce a 0 level output.

gate, OR. Any one input or more than one input having a 1 level signal will produce a 1 level output.

gain bandwidth product. The product of closed-loop gain and its corresponding closed-loop bandwidth. This product is often constant in operational amplifiers.

gain margin. The amount of gain change at 180 degree phase-shift angle frequency that would produce instability.

gain stability. The extent to which the sensitivity of an instrument remains constant with time.

graticule. A scale for measurement of quantities displayed on the cathode-ray tube of an oscilloscope.

ground. A point in a circuit used as a common reference or datum point in measuring voltages; the conducting chassis or framework on which an electrical circuit is physically mounted and to which one point in a circuit is often connected; the earth or a low-resistance conductor connected to the earth at some point and having no potential difference from another conductor connected to the earth at the same point.

ground loop. The generation of undesirable signals within a ground conductor, owing to circulation currents within the conductor which originate from a second source of potential—frequently as a result of connecting two separate grounds to a signal circuit.

guarded input. An input that has a third terminal which is maintained at a potential near the input-terminal potential for a single-ended input—or near the mean input potential for differential input. It is used to shield the entire input circuit.

harmonics. Those frequencies which are integer multiples of the fundamental frequency of a periodic signal.

harmonic distortion. The generation of unwanted harmonics of the input frequency which appear at the output of the system. Total harmonic distortion is the rms sum of all harmonics.

high-pass filter. A filter that transmits alternating current above a given cutoff frequency and substantially attenuates all other currents.

hysteresis. The summation of all effects, under constant environmental conditions, which causes the output of an instrument to assume different values at a given stimulus point when that point is approached first with increasing stimulus and then with decreasing stimulus. The word is most typically applied to the relationship between magnetizing force and magnetic flux density in an iron-core transformer. In the instrumentation field, hysteresis is the same thing as deadband.

hysteresis loop. For a magnetic material in a cyclically magnetized condition, a curve (usually with rectangular coordinates) showing, for each value of the magnetizing force, two values of the magnetic flux density—one when the magnetizing force is increasing, the other when it is decreasing.

impedance. An indication of the total opposition that a circuit or device offers to the flow of alternating current at a particular frequency.

inductive pickup. Signals generated in a circuit or conductor owing to mutual inductance between it and a disturbing source.

inhibit. To prevent an action or acceptance of data by applying an appropriate signal to the appropriate input.

inhibit pulse. A pulse that prevents flux reversal of a magnetic cell by certain specified drive pulses.

input/output buffer. An autonomous storage unit which accumulates blocks of information from the input/output unit (usually several words) for distribution to the computer and consisting of control, storage, and an input/output unit.

input/output unit. The input/output unit consists of three sections—the input/output mechanism, storage to accumulate a convenient amount of information, and control logic. The latter supervises the accumulation and distribution of information from the intermediate medium and the computer or awaiting storage devices.

internal calibration. Calibration by an internal voltage source (provided with the instrument) rather than an external standard.

internal memory. The total memory or storage which is accessible automatically to the computer. This equipment is part of and directly controlled by the computer.

inverter. A circuit whose output is always in the opposite state from the input. Also called a NOT circuit.

inverting amplifier. An amplifier whose output polarity is reversed as compared to its input. Such an amplifier obtains its negative feedback by a connection from output to input and, with high gain, is widely used as an operational amplifier. An operational dc power supply can also be described as a high-gain inverting amplifier.

isolated. Refers to that condition where a conductor, circuit, or device is not only insulated from another (or others), but the two are mutually unable to engender current, emf, or magnetic flux in each other. As commonly used, insulation is associated predominately with dc, whereas isolation implies additionally a bulwark against ac fields.

isolation voltage. A rating for a power supply which species the amount of external voltage that can be connected between any output terminal and ground (the chassis). This rating is important when power supplies are connected in series.

I.T.D.E. (interchannel time displacement error). The timing error, in microseconds, between tracks on the tape caused by dynamic skew measured normally between adjacent tracks on the same head stack or across the head stack.

jitter. An aberration of a repetitive display indicating instability of the signal or of the instrument. May be random or periodic and is usually associated with the time axis.

large-signal characteristics. The characteristics of the amplifier when full (rated) output signals are produced.

linearity. The "straight-lineness" of the transfer curve between an input and an output; that condition prevailing when output is directly proportional to input.

loading error. A loss of output signal from a device owing to a current drawn from its output. It increases the voltage drop across the internal impedance where it would be better to have no voltage drop at all.

loop gain. A measure of the feedback in a closed-loop system, being equal to the ratio of the open-loop to the closed-loop gains. The magnitude of the loop gain determines the error attenuation and, therefore, the performance of an amplifier used as a voltage regulator.

low-pass filter. A filter that transmits alternating current below a given cutoff frequency and substantially attenuates all other currents.

machine cycle. The smallest period of time or complete process of action that repeats itself in order.

magnetic damping. The damping of a mechanical motion by means of the reaction between a magnetic field and the current generated in a conductor moving through that field. The resistance of this conductor converts excess kinetic energy to heat.

magnetic field. Any space or region in which magnetizing forces are of significant magnitude with respect to conditions under consideration. A magnetic field is produced by any current flow, or a permanent magnet including the earth itself.

magnified sweep. A sweep mode of an oscilloscope whose time per division has been decreased by amplification of the sweep wave form rather than by changing the time constants used to generate it.

memory. A device into which information can be introduced and then extracted at a considerably later time.

memory capacity. The amount of information which a memory unit can store. It is often measured in the number of decimal digits or the number of binary digits which the memory unit can store.

multivibrator. An electronic device which may be found in either of two states. It may be observed by examining the state of either of two output connections. There are three kinds of multivibrators. The bistable multivibrator has two input leads; a signal on either causes the device to assume the corresponding output state regardless of its previous state. The monostable multivibrator has but one lead which, when energized, causes the device to assume the corresponding output state (say 1) for a fixed length of time (τ) and then return to the 0 state. The astable multivibrator or free-running multivibrator has no inputs and alternatively assumes its 0 and 1 states, remaining in each for a relatively fixed time. The trigger is a multivibrator whose inputs are connected together and which operates so that an input signal causes the device to assume the state complementary to the one it has just occupied.

noise. Any unwanted electrical disturbance or spurious signal which modified the transmission, measurement, or recording of desired data.

noise bandwidth. The bandwidth of an ideal filter having the same maximum gain and passing the same average power from a white-noise source as the actual network.

noise figure. A measure of the noise in the circuit defined by the ratio of the total noise power per unit bandwidth to the portion of output noise power caused by a source resistance.

noninverting connection. The closed-loop connection when the forward gain is positive for dc signals ($0°$ phase shift).

nonlinearity. The prevailing condition (and the extent of its measurement) under which the input-output relationship fails to be a straight line.

nonstationary process. An ensemble of signals such that one of the averages over the ensemble is not constant with time.

notch filter. An arrangement of electronic components designed to attentuate or reject a specific frequency band with sharp cutoff at either end.

null-balance. A condition of balance in a device or system which results in zero output or zero current input.

octave filter. A constant percentage bandwidth filter in which $f_2 = 2^{n/2}f_0$ and $f_1 = 2^{-n/2}f_0$, where f_2 and f_1 are the upper and lower cutoff frequencies, respectively, $f_0 = \sqrt{f_1 f_2}$ is the center frequency, and n is the number of octaves.

offset. The change in input voltage required to produce a zero output voltage in a linear amplifier circuit. In digital circuits it is the dc voltage on which a signal is impressed.

offset current. A dc current flowing into an amplifier input terminal when the input or reference source of current is disconnected. The fixed or initial part of the offset current may be adjusted to zero by summing a current of opposing direction.

off-ground. The voltage above or below ground at which a device is operated.

offset voltage. A dc potential remaining across an amplifier's input terminals when the output voltage is zero. The fixed or initial part of the offset voltage may be adjusted to zero.

open loop. A tape drive technique which employs no decoupling of the head area from the reels. This technique of tape drive is common to the audio recorder.

open-loop gain. The gain, measured without feedback, is the ratio of the voltage appearing across the output terminal pair to the causative voltage required at the input.

open-loop output impedance. The impedance looking into the output terminal, operating without feedback, and in the linear amplification region.

open-loop voltage gain. The voltage gain of an amplifier operated without feedback.

optimum damping. The value of damping which provides fast response with some overshoot. Optimum damping is about 65% of critical damping.

oscilloscope. An oscillograph primarily intended for the immediate viewing of the graphic plot—most commonly used to denote a cathode-ray oscilloscope.

output impedance. The output impedance is the ratio of the change in output voltage to a change in output (load) current as a function of frequency.

output voltage. The maximum output voltage available when the operational amplifier is operated in the linear amplification region.

overshoot. In the display of a step function (usually of time), that portion of the wave form which, immediately following the step, exceeds its nominal or final amplitude.

parallel. This refers to the technique for handling a binary data word that has more than one bit. All bits are acted upon simultaneously.

parallel operation. The flow of information through the computer or any part of it using two or more lines or channels simultaneously. The connection of two or more elements or power supplies such that corresponding polarity terminals are connected together and their currents sum in a load.

parallel storage. Storage in which all bits, characters, or words are equally accessible in time.

parity check. Use of a digit, called the "parity digit," carried along as a check which is 1 if the total number of ones in the machine word is odd, and 0 if the total number of ones in the machine word is even (odd parity). Even parity uses the reverse conditions.

passive elements.   Resistors, inductors, or capacitors; elements without gain.

passive network.   A network without gain elements.

period.   The time required for a complete oscillation or for a single cycle of events. The reciprocal of frequency.

periodic signal.   A signal in which the amplitude as a function of time repeats itself in its entirety every time interval, or period, T_0. The reciprocal of T_0 is called the fundamental frequency.

phase.   In a periodic function or wave, the fraction of the period which has elapsed measured from some fixed origin. If the time for one period is represented as $360°$ along a time axis, the phase position is called the phase angle.

phase-compensation network.   A network used to provide closed-loop stability.

phase margin.   The additional amount of phase shift of the output signals at the open-loop unity gain crossover frequency that would produce instability.

phase response.   Linear phase response is the ability of a system to pass all frequencies within its passband with the same time delay.

phase shift.   A change in the phase relationship between two periodic functions.

piezoelectric.   Having the property of producing different voltages on different crystal faces when subjected to a stress (compression, tension, twist, etc.) or of producing a stress when subjected to such voltages.

power factor.   The ratio of real to reactive power at the ac line input to an equipment.

precision.   The degree of exactness with which a quantity is stated; a relative term often based on the number of significant digits in a measurement.

probability density.   The probability of finding in one measurement the variable x within dx.

quantizer.   A device which converts an analog quantity into a digital number.

range.   The upper and lower limits between which an instrument's input may be received and for which the instrument is calibrated.

readout, destructive.   When the reading of information in a storage medium destroys the information.

real-time operation.   Solving problems in real time. Also, processing data in time with a physical process so that the results of the data-processing are useful in guiding the physical operation.

recovery time.   A measure of the transient response to a step load change, recovery time signifies the time required for the controlled parameter to return to within a specified level.

rectifier.   A device used to convert an alternating (ac) wave form to a direct current (dc). Half-wave rectifier: produces dc from ac by passing alternate half cycles of the same polarity. Full-wave rectifier: produces dc from ac by reversing alternate half cycles so they are of the same polarity.

register. An interconnection of computer circuitry, made up of a number of storage devices (usually flip-flops) to store a certain number of digits, usually one computer word.

regulation. The process of controlling a parameter (voltage or current) for the purpose of maintaining it constant. The amount of change observed in the controlled parameter versus the causative variation in line, load, temperature, or time.

regulation, load. The process of regulating against changes in load. The amount of output change experienced when the load varies through prescribed limits.

regulation, line. The process of regulating against changes in primary (line) voltage. The amount of output change experienced when the line varies through a prescribed amount.

regulated power supply. An energy converter containing circuits or elements designed to make its output resemble the ideal of either a voltage source (low impedance) or a current source (high impedance). Usually also containing elements intended to make the output insensitive to changes in the source or line.

reliability. The probability that an instrument's repeatability and accuracy will continue to fall within specified limits.

remote error sensing. A means by which a regulator circuit senses the potential at a remote point (usually the load) and causes the power supply to control that voltage. This connection is used to compensate for the voltage drops in connecting wires.

repeatability. The maximum deviation from the mean of corresponding data points taken from repeated tests under identical conditions; the maximum difference in output for any given identically repeated stimulus with no change in other test conditions.

resistance strain gage. A metallic wire or foil grid that produces a resistance change directly proportional to its change in length (strain).

resolution. The smallest change in applied stimulus that will produce a detectable change in the instrument output.

resonant frequency. The first point in the frequency spectrum (other than dc) where a device looks purely resistive. When no damping is present the resonant frequency and the undamped natural frequency are identical. When damping is present the resonant frequency is always lower than the undamped natural frequency.

response time. A measure of the transient response to a step load change.

ringing. An oscillatory transient occurring in the output of a system as a result of a suddenly applied change in input. Usually damped in time.

ripple. An unwanted ac component on the output of a dc supply.

ripple counter. A binary counting system in which flip-flops are connected in series.

rise time. The interval between the instants at which the instantaneous amplitude first reaches specified lower and upper limits, usually 10 and 90% of the nominal or final amplitude of the step. A measure of the time required for the output voltage of a state from a low-voltage level (0) to a high-voltage level (1) once a level change has been started.

rolloff. A gradually increasing loss or attenuation with increase or decrease of frequency beyond the substantially flat portion of the amplitude-frequency response characteristic of a system or transducer.

scale factor. One or more factors used to multiply or divide quantities occurring in a problem and convert them into a desired range.

sensitivity. The property of an instrument which determines scale factor. As commonly used, the word is often short for "maximum sensitivity," or the minimum scale factor with which an instrument is capable of responding.

serial. This refers to the technique for handling a binary data word which has more than one bit. The bits are acted upon one at a time.

serial operation. The flow of information through the computer or in any part of it using only one line or channel at a time. The connection of two or more elements or supplies such that alternate polarity terminals are connected together and their voltages sum in a load. Load current is equal and common through each supply. The extent of series connection is limited by the maximum specified potential rating between any output terminal and ground.

serial storage. Storage in which time is one of the coordinates used to locate any given bit, character, or (especially) word. Storage in which words, within given groups of several words, appear one after the other in time sequence, and in which access time therefore includes a variable latency or waiting time of zero to many word-times, is said to be serial by word. Storage in which the individual bits comprising a word appear in time sequence is serial by bit. Storage for coded-decimal or other nonbinary numbers in which the characters appear in time sequence is serial by character.

shift. To move the characters of a unit of information column-wise right or left. For a number, this is equivalent to multiplying or dividing by a power of the base of notation.

shift pulse. A drive pulse which initiates shifting of characters in a register.

shift register. An arrangement of circuits, specifically flip-flops, which is used to shift serially or in parallel.

shunt calibration. A form of secondary calibration in which a resistor is placed in parallel across one element of a resistive bridge in order to obtain a known and deliberate electrical unbalance.

shunt regulator. A form of control employing the active regulating element in shunt (parallel) with the output.

signal averaging. A signal processing technique which takes advantage of redundant information provided by a repetitive signal. Equivalent to cross-correlation of the repetitive signal with a pulse train.

single-ended (amplifier). An amplifier with one input terminal and one output terminal tied to a common point and therefore operating at a common potential. This point may or may not be connected to ground.

single-point grounding. A grounding system that attempts to confine all return currents to a network which serves as the circuit reference. The phrase single-point grounding does not imply that the grounding system is limited to one earth connection. To be effective, no appreciable current is allowed to flow in the circuit reference, that is, the sum of above return currents is zero.

single sweep. Operating mode for a triggered-sweep oscilloscope in which the sweep must be reset for each operation, thus preventing unwanted multiple display. In the interval after the sweep is reset and before it is triggered it is said to be armed.

slewing rate. The maximum rate of change of output voltage of an operational amplifier when operated in the linear amplification region, in response to a step change in input signal. Rate at which the output can be driven from limit to limit over the dynamic range.

small-signal bandwidth. Same as −3-dB bandwidth.

small-signal characteristics. The characteristics of an operational amplifier operating in the linear amplification region.

stability. Property of retaining defined electrical characteristics for a prescribed period. Independence or freedom from changes in one quantity as the result of a change in another.

stationary process. An ensemble of signals such that all averages over the ensemble are independent of time.

step-function response. The characteristic curve or output plotted against time resulting from the input application of a step-function (a function which is zero for all values of time prior to a certain instant and a constant for all values of time thereafter).

summing point. See null junction.

synchronous. Operation of a switching network by a clock pulse generator. All circuits in the network switch simultaneously. All actions take place synchronously with the clock.

synchronized sweep. A sweep which would free-run in the absence of an applied signal but in the presence of the signal is synchronized by it.

thermistor. A semiconductor whose resistance is extremely temperature-sensitive. Like carbon, thermistors have negative temperature coefficients.

thermocouple. A temperature-sensing device consisting of two dissimilar metal wires joined together at both ends to deliberately incur the Seebeck effect. One wire of the circuit is opened (both output terminals are the same metal), and a small emf appears across the terminals upon application of heat at one junction. The magnitude of this emf is proportional to the difference in temperature between the "measuring junction" (located at the point of measurement) and the "reference junction" (usually located in the measuring instrument).

throughput rate. The rate at which information can be processed.

time-division multiplexing. Multichannel transmission of continuous input signals in sampled (digitized) form whereby the samples from individual channels are transmitted sequentially over a common transmission link. Each sampled channel occupies the entire bandwidth of the transmission link.

toggle. To switch between two states as in a flip-flop.

tracking. Connecting two or more power supplies together so that the positive outputs and negative outputs maintain a fixed relationship, once established.

transducer. A device for translating faithfully the changing magnitude of one kind of quantity into corresponding changes of another kind of quantity. The second quantity often has dimensions different from the first and serves as the source of a useful signal. It is convenient to think of the first kind of quantity as an "input" and the second kind as an "output." There may or may not be significant energy transfer from the transducer's input to output. For example, a photocell transducer does transfer energy; a strain-gage-based transducer does not.

transfer function. The ratio of the output and the input signals (gain or loss) often as a function of frequency.

transient. A phenomenon experiencing a change as a function of time; something which is temporary; a build-up or breakdown in the intensity of a phenomenon until a steady state condition is reached; an aperiodic phenomenon; the time rate of change of energy is finite and some form of energy storage is usually involved.

trigger. A pulse used to initiate some function.

triggering level. The instantaneous level of a triggering signal at which a trigger is to be generated. Also the name of the control which selects the level.

truth table. A chart that tabulates and summarizes all the combinations of possible states of the inputs and outputs of a circuit, and tabulates what will happen at the output for a given input combination.

tuned filter. An arrangement of electronic components which either attenuates signals at a particular frequency and passes signals at other frequencies, or vice versa.

unipolar. Having but one pole, polarity, or direction. Applied to amplifiers or power supplies, it means that the output can vary in only one direction from zero and, therefore, will always contain a dc component.

unity-gain bandwidth. A measure of the gain-frequency product of an amplifier. Unity gain bandwidth is the frequency at which the open-loop gain becomes unity, based on a 6-dB octave crossing.

unity-gain crossover frequency. The frequency at which the open-loop voltage-gain curve crosses through unity gain, or 0 dB.

virtual ground. The condition existing at the input terminals of a high-gain voltage amplifier. If this gain is large, the input is small and may be assumed "virtually zero;" thus the input terminal is at virtually ground potential.

voltage comparator.   An amplifying device with a differential input that will provide an output polarity reversal when one input signal exceeds the other. When operating with open loop and without phase compensation, operational amplifiers make fast and accurate voltage comparators.

watt.   The practical unit for power and therefore the rate at which energy is converted to work or dissipated as heat. In the case of electricity, power in watts equals the product of voltage times in-phase current.

white noise.   Noise whose power spectral density is constant over the frequency range of interest. Also called impulse noise.

word.   An ordered set of characters which has at least one meaning and is stored, transferred, or operated upon by the computer circuits as a unit. A word is treated as an instruction by the control unit and as a numerical quantity by the arithmetic unit.

Zener diode.   A semiconductor used as a constant voltage reference or control element in various electronic circuits—particularly power supplies. Differs from other diodes in that its electrical properties are derived from a rectifying junction which works at a reverse-bias avalanche condition.

APPENDIX A

DECIBELS

It is common for instrumentation systems to increase or decrease the magnitude of a signal a thousand times or more. This change may be performed by one or a series of elements. It has been found useful to express this change on a logarithmic scale called the decibel. The decibel is *not* a unit of measure. It is the logarithm of the ratio of two power, or power-like, quantities. The decibel is defined as

$$dB = 10 \log_{10} \frac{P_2}{P_1} \qquad (A\text{-}1)$$

where P_1 and P_2 are two powers or power-like quantities. From (A-1) it is apparent that when decibel notation is used to express the magnitude of a power on an absolute scale, one of the two powers involved in the ratio must be defined. Thus in (A-1), P_1 is usually defined as the reference quantity.

When dealing with voltage or voltage-like quantities, the decibel must be modified. Thus applying Ohm's law to (A-1) and assuming that the resistances (impedances) are equal gives

$$dB = 20 \log_{10} \frac{V_2}{V_1} \qquad (A\text{-}2)$$

where V_1 and V_2 are two voltages or voltage-like quantities. Again V_1 is usually defined as the reference quantity.

When the gain of an amplifier or the attenuation of an attenuator (or potentiometer) is given, it is assumed that the input quantity is the reference and that the output is so many decibels above or below the implied reference quantity. Table A-1 lists the percentage deviation of voltage and power to decibels for several common values. Table A-2 lists decibels to voltage and power ratios for several common values.

Table A-1 Percentage Deviation of Voltage and Power to Decibels

(−) Minus			(+) Plus	
Voltage (dB)	Power (dB)	%	Voltage (dB)	Power (dB)
0.0087	0.0043	0.1	0.0087	0.0043
0.0174	0.0087	0.2	0.0174	0.0087
0.0261	0.0130	0.3	0.0260	0.0130
0.0348	0.0174	0.4	0.0347	0.0173
0.0435	0.0218	0.5	0.0433	0.0217
0.0523	0.0261	0.6	0.0520	0.0260
0.0610	0.0305	0.7	0.0606	0.0303
0.0698	0.0349	0.8	0.0692	0.0346
0.0785	0.0393	0.9	0.0778	0.0389
0.0783	0.0436	1.0	0.0864	. 0.0432
0.1755	0.0877	2.0	0.1720	0.0860
0.2646	0.1323	3.0	0.2567	0.1284
0.3546	0.1773	4.0	0.3407	0.1703
0.4455	0.2228	5.0	0.4238	0.2119
0.5374	0.2687	6.0	0.5061	0.2531
0.6303	0.3152	7.0	0.5877	0.2938
0.7242	0.3621	8.0	0.6685	0.3342
0.8192	0.4096	9.0	0.7485	0.3743
0.9151	0.4576	10.0	0.8279	0.4139
1.9382	0.9691	20.0	1.5836	0.7918
3.0980	1.5490	30.0	2.2789	1.1394
4.4370	2.2185	40.0	2.9226	1.4613
3.0103	6.0206	50.0	3.5218	1.7609
3.9794	7.9588	60.0	4.0824	2.0412
5.2288	10.4576	70.0	4.6090	2.3045
6.9897	13.9794	80.0	5.1055	2.5527
10.0000	20.0000	90.0	5.5751	2.7875
∞	∞	100.0	6.0206	3.0103

(−) Minus			(+) Plus	
Voltage Ratio	Power Ratio	dB	Voltage Ratio	Power Ratio
1.0000	1.0000	0.00	1.0000	1.0000
0.9886	0.9772	0.1	1.012	1.023
0.9772	0.9550	0.2	1.023	1.047
0.9661	0.9333	0.3	1.035	1.072
0.9550	0.9120	0.4	1.047	1.096
0.9441	0.8913	0.5	1.059	1.122
0.9333	0.8710	0.6	1.072	1.148
0.9226	0.8511	0.7	1.084	1.175
0.9120	0.8318	0.8	1.096	1.202
0.9016	0.8128	0.9	1.109	1.230
0.8913	0.7943	1.0	1.122	1.259
0.8810	0.7762	1.1	1.135	1.288
0.8710	0.7586	1.2	1.148	1.318
0.8610	0.7413	1.3	1.161	1.349
0.8511	0.7244	1.4	1.175	1.380
0.8414	0.7079	1.5	1.189	1.413
0.8318	0.6918	1.6	1.202	1.445
0.8222	0.6761	1.7	1.216	1.479
0.8128	0.6607	1.8	1.230	1.514
0.8035	0.6457	1.9	1.245	1.549
0.7943	0.6310	2.0	1.259	1.585
0.7499	0.5623	2.5	1.334	1.778
0.7079	0.5012	3.0	1.413	1.995
0.6683	0.4467	3.5	1.496	2.239
0.6310	0.3981	4.0	1.585	2.512
0.5957	0.3548	4.5	1.679	2.818
0.5623	0.3162	5.0	1.778	3.162
0.5309	0.2818	5.5	1.884	3.548
0.5012	0.2512	6.0	1.995	3.931
0.4732	0.2239	6.5	2.113	4.467
0.4467	0.1995	7.0	2.239	5.012
0.4217	0.1778	7.5	2.371	5.623
0.3981	0.1585	8.0	2.512	6.310
0.3758	0.1413	8.5	2.661	7.079
0.3548	0.1259	9.0	2.818	7.943
0.3350	0.1122	9.5	2.985	8.913
0.3162	0.1000	10.0	3.162	10.000
0.1000	10^{-2}	20.0	10.000	10^2
0.03162	10^{-3}	30.0	31.620	10^3
0.01	10^{-4}	40.0	100.00	10^4
0.003162	10^{-5}	50.0	316.20	10^5
0.001	10^{-6}	60.0	1,000.00	10^6
0.0003162	10^{-7}	70.0	3,162.00	10^7
0.0001	10^{-8}	80.0	10,000.00	10^8

APPENDIX B

TRANSMISSION CABLES*

An ideal transmission system would transmit a signal without distortion or attenuation. However, it is impossible for a signal to be transmitted without some attenuation. Recall (2-9) where it was shown that for the ideal transmission of a signal through a system, the system had the properties that its frequency response, $A(\omega)$, was a constant and its phase, $\theta(\omega)$, was proportional to frequency; $\theta(\omega) = \omega t_d$. It can be shown that a transmission line with a series resistance per unit length R, series inductance per unit length L, shunt conductance per unit length G, and shunt capacitance per unit length C will transmit signals without distortion if $R/L = C/G$. Furthermore, the time delay is given by $t_d = \sqrt{LC}$ and the attenuation by $A(\omega) = \sqrt{RG}$. It can also be shown that if the line is terminated into a resistive impedance equal to $R_0 = \sqrt{L/C}$, commonly called the characteristic impedance, no reflections will take place when the signal reaches the end of the line.

It is common in microsecond and nanosecond circuitry for commercial instruments to have their output and/or input connections made through matched lines. In this way input and output impedances are made resistive, and it becomes feasible to make interconnections through cables without introducing reflections. The characteristic impedances most common in commercial coaxial cables are nominally 50 ohms and nominally 72 ohms. In general, the 72-ohm cables are used to match the 72-ohm impedance of dipole antennas. The matched line is much the same as impedance matching for maximum power transfer discussed in Section 3-6. Thus if the load resistance is equal to the characteristic impedance of the line, the maximum power is transferred to the load. In such a case there is no reflected power.

In all cases of potential interference, low or high frequency, shielded cable should be used in making interconnections to protect against magnetic and capacitive stray fields. Grounded coaxial cable can be used from 20 kHz to 5 GHz for most systems. But even coaxial if subjected to very strong interference will not protect completely the desired signal. Then more sophisticated cable and equipment isolation techniques must be used depending upon the frequency of the interfering noise and how it enters the cable system.

* Much of the information contained in this and the following appendices was obtained from Applications Booklet No. 101, Brush Instruments Division, Cleveland, Ohio, (November 1969).

What additional measures are taken to reduce noise will reduce outgoing radiation and crosstalk as well.

Coaxial cable consists of an inner and an outer conductor insulated from each other, with both conductors carrying the desired signal currents (source to load and return). Since the outer conductor is usually grounded at the source, load, bulkheads, and other intermediate points, "ground-loop" or "common-mode" currents caused by potential differences of external noise sources are also carried on the outer conductor. Since the desired signal and the undesired noise are both carried on the same outer conductor simultaneously, noise will be introduced into the system, greatly reducing the signal-to-noise ratio. Low-frequency signals (20 kHz to 6MHz) are particularly susceptible to both ground-loop and common-mode interference. In this case, coaxial cable is recommended with the complete coaxial chain having a minimum number of outer conductor ground contacts. Reducing the number of ground connections reduces the number of possible ground loops. Therefore major equipment, relays, switches, connectors, patch panels, and so on, should be isolated from ground with one ground connection at the source as discussed in Apendix C.

Where strong radiated noise fields exist, such as high-powered radar, broadcast stations, power lines, fluorescent lighting, office and industrial machinery, multiple cable runs, and such, the cable conductors act as receiving antennas or secondary windings of transformers and pick up the external noise sources. Particularly bad sources of noise pickup are the "crosstalk" or induced currents encountered in large multiple cable installations. To protect against these radiated noise sources, either triaxial cable or twinaxial cable can be used.

Triaxial cable is coaxial cable with an additional outer copper braid insulated from the signal-carrying conductors that acts as a true shield and protects the enclosed coaxial conductors. This braid or shield is grounded and bypasses both ground loop and capacitive field noise currents away from the signal carrying coaxial, thereby greatly improving the signal-to-noise ratio over standard coaxial cable usage. Triaxial cable is also used in "driven shield" applications where the inner conductor and first shield are driven in parallel at the transmitting end and work against the outer braid which is insulated above ground. At the receiving end, the inner braid is left floating providing a "Faraday" shield between the inner conductor and outer braid. In this way, the cable distributed capacity is greatly reduced thereby reducing cable losses and loading. This application is most effective in low-frequency transducer data systems where the distributed capacity in coaxial cable limits the data capability. Still another use for triaxial is to use the two outer braids as a low-impedance transmission line (approximately 12 ohms) which can be used to carry high-current pulses to low-impedance laser lamps or exploding bridge wire (EBW) ordanance systems. Triaxial cable and connectors completely insulated from ground are available for these applications.

Twinaxial cable is a two-conductor twisted balanced wire line having a specific impedance, with a grounded shielding braid around both wires. Twisting the two balanced signal carrying wires provides cancellation of any random-induced noise voltage pickup, thereby giving protection against magnetic noise field of the low-frequency variety that passes through the copper braid. This cable also provides protection against ground loops and capacitive fields as does triaxial cable. Twinaxial cable usefulness however is limited to approximately 10 MHz since it has high transmission losses above this frequency. Concentric twinaxial cable and connectors are available for low-frequency and video-distribution systems.

Additional common-mode rejection of noise can be obtained in instrumentation systems, where thermocouple and other transducer information must be remotely recorded, by using twinaxial with only one ground contact located at the transducer. Insulated concentric twinaxial connectors are available.

APPENDIX C
GROUND LOOPS

The ground loop is the largest source of electrical noise between electronic modules. More than one ground on a signal circuit or signal cable shield produces a common impedance coupling or ground loop between these two points. This generates large 60-Hz electrical noise currents which are in series and combined with the useful signal. The magnitude of ground loop current is directly proportional to the difference in absolute potential between the two grounds. In most cases a ground loop through either a cable shield or signal circuit will produce so much 60-Hz noise that it will obscure millivolt level signals.

Two separate grounds are seldom, if ever, at the same absolute voltage. This potential difference creates unwanted current in series with one of the signal leads. In Figure C-1 the potential difference between earth ground 1 and earth ground 2 produces ground-loop current in the lower signal lead from the signal source to the input of the amplifier, causing ground-loop noise to be combined with the useful signal. There is a second ground loop in Figure C-1 through the signal cable shield from the signal source to the

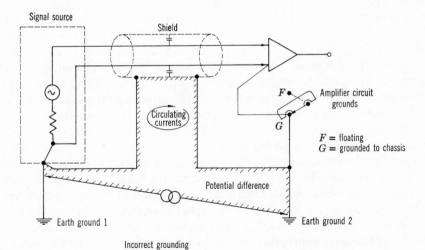

Incorrect grounding

Figure C-1 Typical ground loop created by more than one ground on a signal circuit.

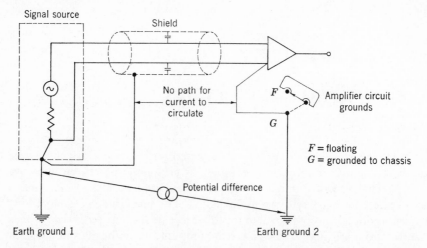

Figure C-2 Elimination of ground loop by turning amplifier input to floating and removing jumper from earth ground.

amplifier. The ground-loop current in the shield is coupled to the signal pair through the distributed capacity in the signal cable. This current is returned through the output impedance of the signal source and back to earth ground 1, adding a second source of noise to the useful signal. Either one of these ground loops is capable of generating a noise signal that is at least 100 times larger than a typical millivolt level signal.

The amplifier shown in Figure C-2 is capable of being floated a few volts off ground. The ground loop through the signal lead can be broken by simply lifting the amplifier grounding strap. The amplifier enclosure is still solidly grounded to earth ground 2, but this will not create a ground loop, since the amplifier enclosure is insulated from the signal circuit. The ground loop through the signal cable shield is eliminated by removing the jumper from the cable shield to earth ground 2. Now the signal source and the signal cable shield are grounded only at the signal source, which is the proper configuration for minimum noise pickup.

In off-ground measurements, the signal cable shield should not be grounded. Effective shielding is secured by stabilizing the signal cable shield with respect to the useful signal. The signal cable should be connected to either the center tap or the low side of the signal source. Since the signal cable shield is being driven by an off-ground common-mode voltage, it is necessary that the cable have appropriate insulation between the shield and the outside of the cable.

APPENDIX D
TYPES OF SIGNAL SOURCES

Transducer or signal source selection is the first step in most measurement or recording applications. Low-impedance devices are preferred, to reduce system noise and to minimize the shunting effect (loading) which the measuring instrument imposes on the source.

The signal source must be properly identified so that it can be matched with an appropriate amplifier. Signal sources fall into six classes, according to the configuration of the output circuit. These are summarized as follows (see Table D-1):

1. A single-ended grounded signal source has two output terminals, one of which is connected to source ground as shown. The ac line-powered signal generator with a two-terminal grounded output is typical.

2. A single-ended floating signal source has two output terminals which are isolated from ground. A floating output can be grounded or reversed without disturbing the circuit. The dry cell battery, the output from a magnetic head, or a two-terminal battery-powered signal generator are typical examples.

3. A single-ended driven-off-ground signal source has two output terminals which are driven off ground by a second voltage. A driven-off-ground signal source can never be grounded. A resistive shunt installed in the hot side of a power line or dc bus for measuring current is a classic example.

4. A balanced grounded signal source has two active output terminals which have equal impedance to a common ground. The output terminals can be reversed without disturbing the circuit. A four-arm wheatstone bridge output that is excited from a grounded power supply is a good example.

5. A balanced floating signal source is one that has two active output terminals which have equal impedance to common point that is floating. A four-arm wheatsone bridge output that is excited from a floating power supply or a center-tapped transformer secondary are typical examples. The output terminals can be reversed or the common terminal can be grounded without disturbing the circuit.

6. A balanced driven-off ground signal source has two active output terminals which have equal impedance to a common point which is driven off ground by a second voltage. The active output terminals can be reversed to invert signal polarity, but it can never be grounded without disturbing or destroying the signal source. An example is a differential output amplifier

Table D-1 Amplifier Compatibility Summary (X = compatible).

Preamp Input Configuration \ Source Configuration	Single-Ended: Grounded	Single-Ended: Floating	Single-ended: Driven-Off Ground	Balanced: Grounded	Balanced: Floating	Balanced: Driven-Off Ground
Single-ended: Grounded		X				
Balanced to ground				X	X	
Single-ended: floating and shielded	X	X	X		X	
Balanced: floating, guarded	X	X	X	X	X	X

which produces an ac signal of ± 30 volts, but operates at about $+60$ volts dc off ground.

If an electrical schematic is not available, a signal source may be identified by using an oscilloscope plus an ohmmeter. To identify a two-terminal source, the ground or low-level input terminal is connected to a good solid ground at the signal source. With the source turned on, the probe of the oscilloscope input is connected to one output terminal of the signal source and then the other. The amplitude and character of these two measurements provide the required information about the source.

A zero signal from one signal-source terminal and a usable signal from the other indicate a single-ended grounded source. Equal 60-Hz noise signals from both terminals indicate a floating source, and if there are only two terminals a floating source is probably single-ended floating. A resistance of several hundred megohms from each terminal to ground confirms that the source is floating. Usable but unequal signals from the two terminals indicate a driven-off ground output. The average of the two signals is the off-ground or common-mode voltage; the difference is the signal amplitude. If the two-terminal source is turned off and the ohmmeter shows unequal resistance from the terminals to ground, the source is probably single-ended driven-off ground. For all three source types, an ohmmeter across the two terminals with the source turned off indicates the source resistance.

To identify a three-terminal signal source, the ground or low-level input terminal is again grounded at the signal source. The probe of the oscilloscope terminal is connected to each of the three-source output terminals in sequence.

A zero signal from one terminal and equal or similar signals from the other two indicates a balanced grounded source. With the signal source turned off, resistance readings from ground to each terminal identify the ground terminal and the source resistance of each active output terminal to ground. Equal 60-Hz noise signals from the three terminals and a resistance of several hundred megohms between all three terminals to ground indicate a balanced floating source. With the source turned off, ohmmeter readings across the three output terminals reveal the common terminal, the source resistance of each leg to common, and the total source resistance.

A usable signal from one terminal to ground and nearly equal signals from the other two indicate a driven-off ground source. The terminal with the minimum signal or a signal different from the other two is probably the common terminal, and the signal from it to ground is the off-ground or common-mode voltage. With the source turned off, equal resistance readings to ground from any pair of terminals and a smaller resistance to ground from a third terminal confirms that the source is balanced driven-off ground. Resistance readings taken across the three output terminals reveals the common terminal, the output resistance of each leg to common, and the total source resistance.

APPENDIX E

AMPLIFIER INPUT CONFIGURATIONS

The single-ended grounded amplifier has two input terminals, one of which is common with the output. This direct connection from the input to output is normally grounded through the third conductor in the ac power cord. The amplifier enclosure is usually internally connected to this common point (see Table D-1).

This type of amplifier works well with balanced floating-signal sources or single-ended floating-signal sources, but it will not operate properly with any type of grounded signal source unless it is connected to the same ground as the amplifier, or unless one of the grounds is disconnected. Any difference in potential between the signal source ground and the amplifier ground will add to or subtract from the true value of the signal being measured. This erroneous noise signal is generally referred to as a "ground loop" and must be avoided.

The balanced-to-ground amplifier has two active input terminals which have equal resistance to a common-ground terminal. This common-input terminal is grounded firmly through the third wire in the ac power cord. Both the enclosure and the low side of the output are grounded normally to this same point inside the amplifier case. It can be operated as a "balanced to-common" amplifier by removing the ground connection in the ac power cord. In this case the amplifier would normally be connected to one ground at the signal source. It can also be operated as a "single-ended grounded" amplifier by simply installing a ground strap from one active input terminal to ground.

The major limitation of a "balanced-to-ground" amplifier is its restricted off-ground voltage capability which decreases as the input attenuator is advanced to a more sensitive position. This type of amplifier will work well with a single-ended floating-signal source, a balanced floating signal source and in most cases with a balanced grounded-signal source.

The single-ended floating amplifier has two input terminals that are isolated electrically from the output terminals. It is normally provided with an internal floating shield which is connected internally to the low side of the input. Both input terminals are free to float up or down in compliance with any common-mode voltage that may appear at the signal source but the

337

capacity to ground of the low input terminal is greater than the capacity to ground of its "hot" input terminal. It is therefore important to connect this amplifier so that any common-mode voltage at the signal source is used to drive the low side of the input. The amplifier enclosure and the low side of the output are grounded normally through the third wire in the ac power cord. This type of amplifier works well with single-ended grounded-signal sources, single-ended floating sources, or single-ended driven-off ground sources and can be used down to millivolt levels with balanced floating-signal sources.

The balanced, floating, and guarded amplifier is the most sophisticated for all dc amplifier types and can be used with all types of self-generating signal sources. Both input terminals are isolated from amplifier chassis and isolated from the output. The input terminals also have equal impedance to a third terminal called the guard shield or simply "guard," which is a full floating internal shield. The guard is used to minimize internal capacity from signal input terminals to chassis ground and to improve the ac common-mode rejection of the amplifier.

Both amplifier input terminals are free to float up or down in compliance with any common-mode voltage that may appear at the signal source. Since both input terminals are floating and have very low capacity to chassis ground, the incoming signals may be grounded, floating, or driven off-ground without affecting accuracy or system noise. When the guard shield is properly connected, the ac noise-rejection characteristics are quite good, so this amplifier can be used over a wide range of signal amplitudes, down to and including the microvolt level. The guard shield is not connected internally, but it is brought out to separate terminals in the amplifier input connector so that it may be connected properly for all types of signal sources.

INDEX